Aluminium Alloys - New Trends in Fabrication and Applications

Aluminium Alloys - New Trends in Fabrication and Applications

Editor

Leroy Sidney

scitus
academics

Aluminium Alloys - New Trends in Fabrication and Applications
Edited by **Leroy Sidney**

ISBN: 978-1-68117-312-2
Library of Congress Control Number: 2016937621

SCITUS Academics LLC,
www.scitusacademics.com
Box No. 4766, 616 Corporate Way,
Suite 2, Valley Cottage,
NY 10989

Printed in United States of America on Acid Free Paper

Preface

Aluminum is the world's most abundant metal and is the third most common element representing eight percent of the earth's crust. Aluminum's versatility makes it the most widely used steel metal. Aluminum compounds have been used for thousands of years, but for the first time some 170 years ago aluminum metal was produced. Aluminum alloys underwent dramatic transformations in extrusion, machining, welding, heat treatment, ultra-fine particle structural changes and increased corrosion resistance. These alloys have therefore quickly gained familiarity in the automotive and space sectors. Aluminum alloys are predominantly aluminum (Al) metal alloys.

This book highlights future aluminum alloy innovations as well as applications in several fields. Aluminum alloys are widely used in structures and components in which light weight or corrosion resistance is necessary. The characteristics of the various aluminum alloys have led to the use of aluminum in industries as diverse as transport, food preparation, and generation of energy, packaging, architecture and electricity. Depending on the application, aluminum may be used to replace other materials, such as copper, steel, zinc, stainless steel, titanium, wood, paper, concrete and composites. The new role of aluminum alloys in photovoltaic cells and concentrated solar energy has been comprehensively described in the context of corrosion and the aggressive environment to which they can be exposed.

Table of Contents

Chapter 1 Durability and Corrosion of Aluminium and Its Alloys: Overview, Property Space, Techniques and Developments 1

Chapter 2 Equal-Channel Angular Pressing and Creep in Ultrafine-Grained Aluminium and Its Alloys 63

Chapter 3 Fabrication of Corrosion Resistance Micro-Nanostructured Superhydrophobic Anodized Aluminum in a One-Step Electrodeposition Process 115

Chapter 4 Influence of Post Weld Heat Treatment on Strength of Three Aluminum Alloys Used in Light Poles 127

Chapter 5 Effects of Cryogenic Forging and Anodization on the Mechanical Properties and Corrosion Resistance of AA6066–T6 Aluminum Alloys 151

Chapter 6 Effects of Deep Rolling and Its Modification on Fatigue Performance of Aluminium Alloy AA6110 169

Chapter 7 Phase Evolution and Mechanical Behavior of the Semi-Solid SIMA Processed 7075 Aluminum Alloy 187

Chapter 8 Aluminium Alloys in Solar Power – Benefits and Limitations 223

Index 265

CHAPTER 1

Durability and Corrosion of Aluminium and Its Alloys: Overview, Property Space, Techniques and Developments

N. L. Sukiman[1, 6], X. Zhou[1], N. Birbilis[1], A.E. Hughes[2], J. M. C. Mol[3], S. J. Garcia[4], X. Zhou[5] and G. E. Thompson[5]

[1] Department of Materials Engineering, Monash University, Clayton, Australia

[2] CSIRO Materials Science and Technology, Melbourne, Australia

[3] TU Delft, Materials Department, Delft, Netherlands

[4] TU Delft, Aerospace Engineering, Delft, Netherlands

[5] School of Materials, The University of Manchester, Manchester, United Kingdom

[6] Department of Mechanical Engineering, University of Malaya, Kuala Lumpur, Malaysia

1. INTRODUCTION

Aluminium (Al) is an important structural engineering material, its usage ranking only behind ferrous alloys (Birbilis, Muster et al. 2011). The growth in usage and production of Al continues to increase (Davis 1999). The extensive use of Al lies in its strength:density ratio, toughness, and to some degree, its corrosion resistance. From a corrosion perspective, which is most relevant to this chapter, Al has been a successful metal used in a number of applications from commodity roles, to structural components of aircraft. A number of Al alloys can be satisfactorily deployed in environmental/atmospheric conditions in their conventional form, leaving the corrosion protection industry to focus on market needs in more demanding applications (such as those which require coating systems, for example, the aerospace industry).

Relatively pure aluminium presents good corrosion resistance due to the formation of a barrier oxide film that is bonded strongly to its surface (passive layer) and, that if damaged, re-forms immediately in most environments; i.e. re-passivation (Davis 1999). This protective oxide layer is especially stable in near-neutral solutions of most non-halide salts leading to excellent pitting resistance. Nevertheless, in open air solutions containing halide ions, with Cl^- being the most common, aluminium is susceptible to pitting corrosion. This process occurs, because in the presence of oxygen, the metal is readily polarised to its pitting potential, and because chlorides contribute to the formation of soluble chlorinated aluminium (hydr)oxide which interferes with the formation of a stable oxide on the aluminium surface.

Aluminium and its alloys readily oxidises, including when Al is present in either in solid solution or in intermetallic (IM) particles. Industrial alloy surfaces however, tend to be as heterogeneous as their underlying microstructures. The surface of a wrought or cast alloy is likely to contain not only aluminium oxide alone, but may for example contain a fragment of a mixed Al-Mg oxide for alloys rich in Mg (Harvey, Hughes et al. 2008)). This is primarily because of the heat of segregation of Mg is high and it has a favorable free energy for oxide formation. If however an Al surface is mechanically undisturbed - then the surface oxide is relatively protective. Though, most real surfaces have some sort of mechanical finishing which results in the formation of a near surface deformed layer (NSDL) and shingling. Shingling occurs where the alloy matrix is spread across the surface including IM particles in abrasion and milling (Scholes, Furman et al. 2006; Muster, Hughes et al. 2009). This is because the IM particles are harder than the surrounding matrix and less susceptible to deformation (Zhou, Liu et al. 2011). Even on polished surfaces, the matrix and the IM particles rapidly form different oxide structures (Juffs, Hughes et al. 2001; Juffs, Hughes et al. 2002). This is almost certainly due to different chemical environments and different electrochemical reactions over the IM particles compared to the matrix. Furthermore, the morphology and the oxide are not continuous from the IM particles to the matrix and this represents a potential defect site in the context of corrosion. For the purposes of descriptions herein, IM particles can be classified into three main types; i) precipitates, ii) constituent particles and iii) dispersoids. Precipitates are typically in the shape of needles, laths, plates or spherical with the size ranging from Angstroms to fractions of a micrometer. They are formed by nucleation and growth from a supersaturated solid solution during low temperature aging and may be concentrated along the grain boundaries. Constituent particles however, are relatively large with irregular shape and the size can be up to

10 micrometers. This type of particle forms during solidification of the alloy and is not fully dissolved by subsequent thermomechanical processing (including solution heat treatments). They can be found in colonies of several IM crystals or different compound types. On the contrary, dispersoids are small particles with size ranging from 0.05 to 0.5 micrometers. They are thermally stable intermetallics of a fine size that are functional for controlling grain size and recrystallisation behavior. Dispersoids form by low level additions of highly insoluble elements such as Cr, Mn or Zr.

This chapter will aim to cover some of the important aspects related to the corrosion of Al-alloys, bearing in mind the role of alloy chemistry. In addition, some of the topical aspects related to techniques and ongoing developments in the general field of Al-alloy corrosion are presented. An attempt has been made to give the reader an overview of the key technical aspects, but unfortunately for comprehensive insight into the topic overall, the size of this chapter alone cannot be a replacement to dedicated monographs on the specific topics at hand; nor the ever-evolving journal literature that represents the state of the art. To aid in the transfer of information, this chapter has been divided into a number of sections to treat the widely varying topics independently.

1.1. The General Performance of the Al-Alloy Classes

The corrosion potential of an aluminium alloy in a given environment is primarily determined by the composition of the aluminium rich solid solution, which constitutes the predominant volume fraction and area fraction of the alloy microstructure (Davis 1999). While the potential is not affected significantly by second phase IM particles of microscopic size, these particles frequently have corrosion potentials (when measured in isolation) differing from that of the solid solution matrix resulting in local (micro-) galvanic cells, when IMs are polarised to the corrosion potential of the alloy. The result is that local currents on the alloy surface differ, establishing anodes and cathodes. Since most of the commercial aluminium alloys contain additions of more than one type of alloying element, the effects of multiple elements on solution potential are approximately additive. The amounts retained in solid solution, particularly for more highly alloyed compositions, depend on production and thermal processing so that the heat treatment and other processing variables influence the final electrode potential of the product.

By measuring the potentials of grain boundaries and grain bodies separately, the difference in potential responsible for local types of corrosion such as intergranular corrosion, exfoliation, and stress corrosion cracking (SCC) can be quantified (Guillaumin and Mankowski 1998; Zhang and Frankel 2003).

By measuring the corrosion potential of IMs (Buchheit 1995), and indeed by measurement of the polarisation response of IMs, even more significant insights into localised corrosion can be gained (Birbilis and Buchheit 2005). Such specialist topics are not dealt with in their entirety herein, however an abridged written summary of the performance of the key Al-alloy classes (as outlined by the Aluminium Association (Hatch 1984)) is provided below.

1.1.1. 1xxx Series Alloys

Corrosion resistance of aluminium increases with increasing metal purity, however the use of the >99.8% grades is usually confined to those applications where very high corrosion resistance or ductility is required. In regards to such specialist applications however, the actual number of applications are very few. Consequently 1xxx series alloys are not commonly used or sold (but do serve as important feedstock to secondary alloy producers or production). In the instance where general-purpose alloys for lightly stressed applications are required, such alloys are approximately 99% pure aluminium and offer adequate corrosion resistance in near neutral environments. 1xxx is also sometimes used in cladding for example AA1230 is used as clad on AA2024 (Hatch 1984)

1.1.2. 2xxx Series Alloys

Copper is one of the most common alloying additions - since it has appreciable solubility and can impart significant strengthening by promotion of age hardening (in fact, the Al-Cu system was the classical/original age hardening system (Hatch 1984)). These alloys were the foundation of the modern aerospace construction industry and, for example AA2024 (Al-4.4Cu-1.5Mg-0.8Mn), which is still used in many applications to this day, can achieve strengths in excess of 500MPa depending on temper (Polmear 2006).

1.1.3. 3xxx Series Alloys

The 3xxx series alloys are a commodity product that is nominally available in the form of thin sheet (for beverage can usage). The key alloying element, Manganese, has a relatively low solubility in aluminium but can improve corrosion resistance when remaining in solid solution. Additions of manganese of up to about 1% form the basis of the non-heat treatable wrought alloys with good corrosion resistance, moderate strength (i.e. AA3003 tensile strength ~110MPa) and exceptionally high formability (Polmear 2006).

1.1.4. 5xxx Series Alloys

Magnesium has significant solubility in aluminium and imparts substantial solid solution strengthening (which can also contribute to enhanced work hardening characteristics) (Davis 1999; Polmear 2006). The 5xxx series alloys (containing <~6% Mg) do not age harden. Whilst it is possible for β-phase (Mg_2Al_3) to precipitate in systems with above ~3%Mg, the β-phase is not a strengthening precipitate and actually weakens the alloy (by depleting the solute of Mg). Nominally, the corrosion resistance of 5xxx series alloys is good and their mechanical properties make them ideally suited for structural use in aggressive conditions (such as marine vessels). Fully work-hardened AA5456 (Al-4.7Mg-0.7Mn-0.12Cr) has a tensile strength of >380MPa. One corrosion issue with fully work-hardened 5xxx series alloys is that the heavy dislocation density (and supersaturation of the solid solution with Mg) can permit the sensitization of the microstructure by precipitation of deleterious β-phase (Mg_2Al_3) during sustained high temperature exposure (i.e. in service) (Baer, Windisch et al. 2000; Searles, Gouma et al. 2002; Davenport, Yuan et al. 2006; Goswami, Spanos et al. 2010).

1.1.5. 6xxx Series Alloys

Silicon additions alone can lower the melting point of aluminium whilst simultaneously increasing fluidity (which is why the vast majority of cast Al products contain various amounts of Si). These alloys are increasing in importance in automotive applications for engine and drive train components – however are yet to realise the majority of market share. Heat-treatable Al-Mg-Si are predominantly structural materials (strengths >300MPa are possible), all of which have an appreciable resistance to corrosion, immunity to SCC and are weldable. To date, 6xxx series alloys are mainly used in extruded form, although increasing amounts of sheet are being produced (Birbilis, Muster et al. 2011). Magnesium and silicon additions are made in balanced amounts to form quasi-binary Al-Mg_2Si alloys, or excess silicon additions are made beyond the level required to form Mg_2Si. Alloys containing magnesium and silicon in excess of 1.4% develop higher strength upon aging.

1.1.6. 7xxx Series Alloys

The Al-Zn-Mg alloy system provides a range of commercial compositions, primarily where strength is the key requirement (and this can be achieved without relatively high cost or complex alloying). Al-Zn-Mg-Cu alloys have traditionally offered the greatest potential for age hardening and as early as 1917 a tensile strength of 580MPa was achieved, however, such alloys were not suitable for commercial use until their high susceptibility to stress corrosion cracking could be moderated (Song, Dietzel et al. 2004; Birbilis, Cavanaugh et al. 2006; Lin, Liao et al. 2006; Lynch, Knight et al. 2009). Aerospace needs led to the introduction of a range of high strength aerospace alloys of which AA7075 (Al-5.6Zn-2.5Mg-1.6Cu-0.4Si-0.5Fe-0.3Mn-0.2Cr-0.2Ti) is perhaps the most well-known, and which is now essentially wholly superseded by AA7150 (or the 7x50 family). The high strength 7xxx series alloys derive their strength from the precipitation of η-phase ($MgZn_2$) and its precursor forms. The heat treatment of the 7xxx series alloys is complex, involving a range of heat treatments that have been developed to balance strength and stress corrosion cracking performance - including secondary (or more) heat treatments that can include retrogression and re-aging (Fleck, Calleros et al. 2000; Ferrer, Koul et al. 2003; Zieliński, Chrzanowski et al. 2004; Marlaud, Deschamps et al. 2010).

1.1.7. 8xxx Series Alloys

Nominally reserved for the sundry alloys, 8xxx series alloys include a number of Lithium (Li) containing alloys. Li is soluble in aluminium to ~ 4 wt% (corresponding to ~ 16 at%). As these alloys of high specific strength and stiffness readily respond to heat treatment, research and development has intensified due to their potential for widespread usage in aerospace applications (Lavernia and Grant 1987; Dorward and Pritchett 1988; Giummarra, Thomas et al. 2007). Based on the impressive lightweight of such alloys, present day aircraft are comprised of some portion of Al-Li based alloys (modern generations of which actually include low Li levels and hence are nowadays designated as 2xxx alloys (Ambat and Dwarakadasa 1992; Garrard 1994; Semenov 2001; Giummarra, Thomas et al. 2007). First generation Li-containing alloys displayed some of the highest corrosion rates of all aluminium alloys, where susceptibility to intergranular corrosion was challenging. Modern Al-Cu-Li seem to have overcome this challenge; however it is also important to recognise that production requires specialised melting and casting, not presently available in most commercial facilities.

2. CORROSION OF ALUMINIUM AND ITS ALLOYS IN AQUEOUS ENVIRONMENT

2.1. Environmental Corrosion of Aluminium

Corrosion in aluminium alloys is generally of a local nature, because of the separation of anodic and cathodic reactions and solution resistance limiting the galvanic cell size. The basic anodic reaction is metal dissolution:

$$Al \rightarrow Al^{3+} + 3e^-$$

While the cathodic reactions are oxygen reduction:

$$O_2 + 2H_2O + 4e^- \rightarrow 4OH^-$$

or hydrogen reduction in acidified solution such as in a pit environment as a result of aluminium ion hydrolysis:

$$2H^+ + 2e \rightarrow H_2$$

It is the interaction between local cathodes and anodes and the alloy matrix that leads to nearly all forms of corrosion in aluminium alloys. These include pitting corrosion, selective dissolution, trenching, intermetallic particle etchout, intergranular attack and exfoliation corrosion. Surface and subsurface grain etchout is also influenced by grain energy which is derived from grain defect density. Grain etchout, has a significant role in exfoliation corrosion since the volume of hydrated aluminium oxide generated during dissolution is larger than the original volume of the grain.

The general consensus for Al and its alloys is that they are resistant towards corrosion in mildly aggressive aqueous environments. The protective oxide layer represents the thermodynamic stability of Al alloys in corrosive environment - acting as a physical barrier as well as capable to repair itself in oxidizing environments if damaged. While the passive layer breakdown mechanism by chloride ions is still in debate (Sato 1990; McCafferty 2010) due to the complexity of the process (Szklarska-Smialowska 2002), the general consensus is that localized attack starts by adsorption of aggressive anions and formation of soluble transitional complexes with the cations at the oxide surface. Thermodynamic principles to explain and predict the passivity phenomenon that controls the corrosion behavior of Al are summarised by Pourbaix-type analysis. This results in a plot of potential vs. pH based on the electrochemical reaction of the species involved, the representation known as a Pourbaix diagram (Pourbaix 1974) as shown in Figure 1.

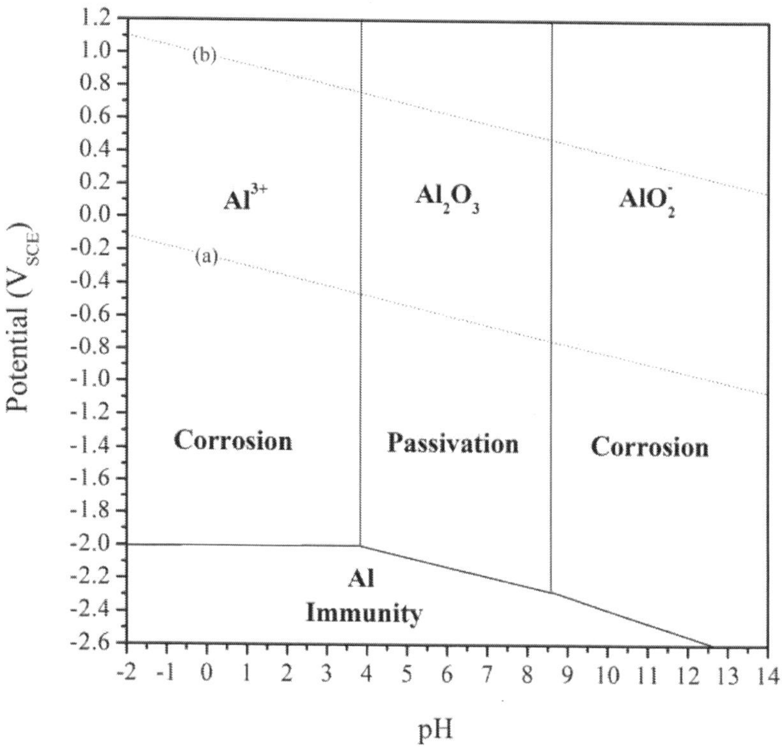

Figure 1. E-pH diagram for pure Al at 25°C in aqueous solution (adapted from Pourbaix 1974). The lines (a) and (b) correspond to water stability and its decomposed product.

It is seen that Al is nominally passive in the pH range of ~4 to 9 due to the presence of an Al_2O_3 film. In environments that deviate from the near neutral range, the continuity of this film can be disrupted in which the film becomes soluble, facilitating the relatively rapid of dissolution the alloy. In the acidic range, Al is oxidised by forming Al^{3+}, whilst AlO^{2-} occurs in alkaline range.

The E-pH diagram gives an impression that corrosion prediction is a straightforward process, however in actual engineering applications, there are several variables that weren't considered by Pourbaix. These include (i) the presence of alloying elements in most engineering metals (ii) the presence of substances in the electrolyte such as chloride (albeit that this has been addressed in more modern computations), (iii) the operating temperature of the alloy, (iv) the mode of corrosion, and (v) the rate of reaction. Taking these factors into account is nominally done on a case by case (i.e. alloy by alloy) basis, and a revised version of an E-pH diagram for 5xxx series alloys in 0.5M sodium chloride is given in Figure 2.

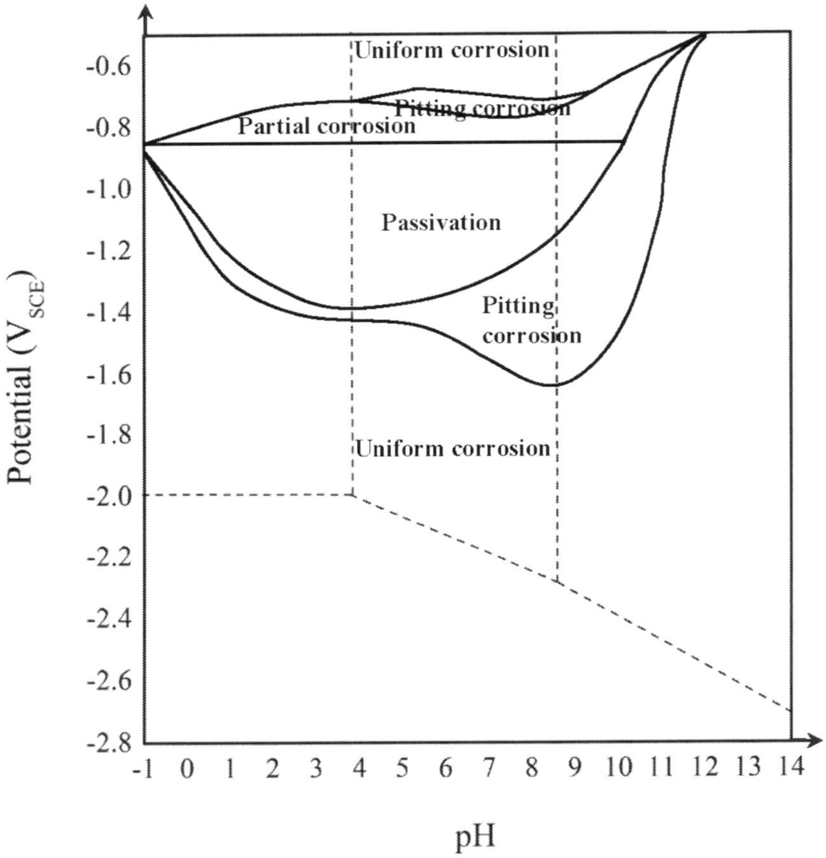

Figure 2. Mode of corrosion based on experimental data for AA5086 in the presence of 0.5M sodium chloride (adapted from Gimenez, Rameau et al. 1981)

Figure 2 indicates windows where localized attack is highly possible in the supposed passive region (Gimenez, Rameau et al. 1981). It is also seen that localised attack is possible across the whole range of pH depending on the specific potential. One should therefore not rely solely on the Pourbaix diagram as a direct index to actual corrosion rates, with rates needing to be independently measured for a given alloy-electrolyte combination (Ambat and Dwarakadasa 1992). Finally, whilst not to be discussed in detail here, it is prudent to indicate that effectively all Al-alloys do not attain practical/empirical immunity as evidence in Figure 1. Cathodic polarisation tends to contribute to alloy deterioration by two modes. Firstly, the accumulation of hydroxyl ions at the Al-surface will cause chemical dissolution of the Al. Secondly, Al is a very strong hydride former, and hydrogen from the cathodic reaction at such negative potentials will serve combine with Al to form hydrides (Perrault 1979).

2.2. Kinetic Stability of Aluminium Alloys

Kinetics represents the rate of reaction during corrosion. When exposed to an aqueous environment, metals stabilise to a value of electrochemical potential that is characteristic of the material and its composition for a given electrolyte. This potential is the potential at which anodic and cathodic reactions upon the metal surface are equal, and the value of this potential is thus significantly influenced by factors that can alter the relative rates of anodic or cathodic reaction efficiency upon the metal surface (i.e. alloying, precipitate state, etc.).

Most typically, the potentiodynamic polarisation test is used to characterise the corrosion performance of an alloy (as far as determination of mechanistic aspects from an instantaneous test). This method gives vital kinetic information such as current density over a range of potentials, pitting potential (if it exists), corrosion potential, the passive current density and potentially more information in reverse scans, etc. Thus factors affecting corrosion as discussed in the previous sections can be evaluated with much higher confidence. For example, Figure 3 shows a polarisation curve of Pure Al compare to AA2024-T3 (Al-4.3Cu-1.5Mg-0.6Mn) in 0.1M NaCl.

Comparison of alloy behavior and dominant reactions can be made in a quantitative manner. The anodic branch of the polarisation curve gives information related to the anodic/dissolution reaction while the cathodic branch represents the reduction reaction (nominally oxygen reduction, but at lower potentials or in strong acids, water reduction). The addition of more noble alloying elements typically increases the corrosion potential to more noble values (Davis 1999) and this is dramatically observed inFigure 3. This ennoblement however does not correlate to the rate of corrosion (as judged by Figure 3), whereby we see the pure Al versus AA2024-T3 has a difference in potential of ~0.5V. In addition, the main practical threat for Al alloys is localised attack such as pitting, so in that vein, a more noble value of pitting potential does not necessarily signify a better corrosion resistance (Frankel 1998;Birbilis and Buchheit 2005). Rather generally, the electrochemical reactions upon Al-alloys are heavily influenced by the chemistry and microstructure of the alloy – which we attempt to discuss in the following section.

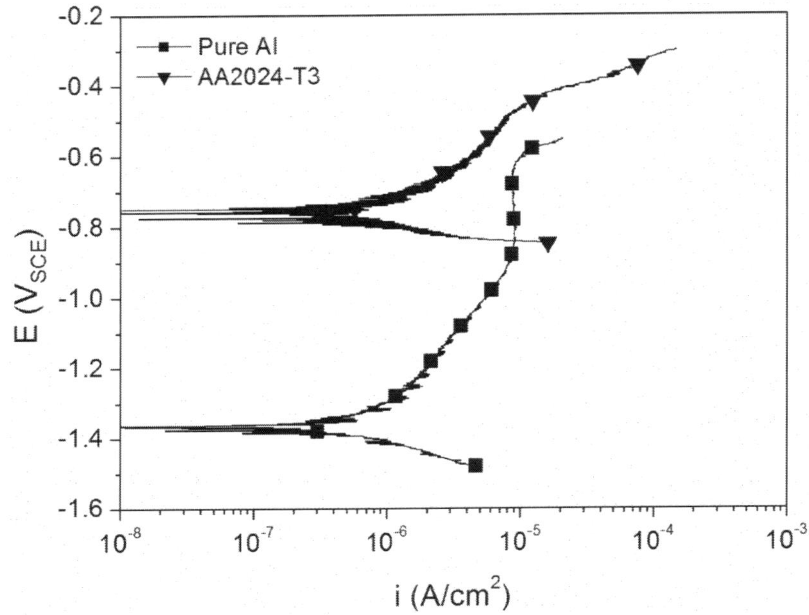

Figure 3. Polarisation curve of pure Al and AA2024-T3 exposed to 0.1M NaCl for 7 days collected at $1mV/s^{-1}$ (adapted from (Sukiman, Birbilis et al. 2010))

Moving beyond potentiodynamic polarisation towards a true measure of kinetic stability in the E-pH domain (similar in concept to Pourbaix diagrams however which give 'speed' and not just thermodynamic likelihood) there are tests which can be done in this regard. In order to develop an improved understanding of overall kinetic stability of a metal over the potential-pH space, methods including the staircase potentio-electrochemical impedance spectroscopy (SPEIS) can be used to establish so-called kinetic stability diagrams, as previously demonstrated for pure Al (Zhou, Birbilis et al. 2010) and depicted in Figure 4. The specifics of SPEIS will be introduced in section 5.3.

In Figure 4, presentation of the reciprocal of polarisation resistance ($1/R_P$) is the metric of reaction 'speed', as it is proportional to the reaction rate at a given condition. The influence of potential and pH is presented not only for pure Al in the sample which was single crystal, but also for a polycrystalline sample. As a result, one is also able to assess the effect of grain structure from Figure 4. Most importantly however, Figure 4 reveals that the rate of reaction stays in a low range in areas that extend beyond that of the Pourbaix diagram in terms of certain E-pH combinations – indicating that although Al may be in a thermodynamically unstable region, the reaction rate can be maintained to be low enough to make it still be useful in an engineering context. Similarly, there are regions of high potential where pure aluminium may be in a thermodynamically stable region, but unusable – owing to transpassive dissolution. Finally, in terms of microstructure effects, it is seen that the exact same material can have a different kinetic response based on structural factors alone. Such differences are not detectible or predicted from thermodynamic analysis in any way, and this highlights the importance of approaches which provide kinetic information to meet demands of engineering applications.

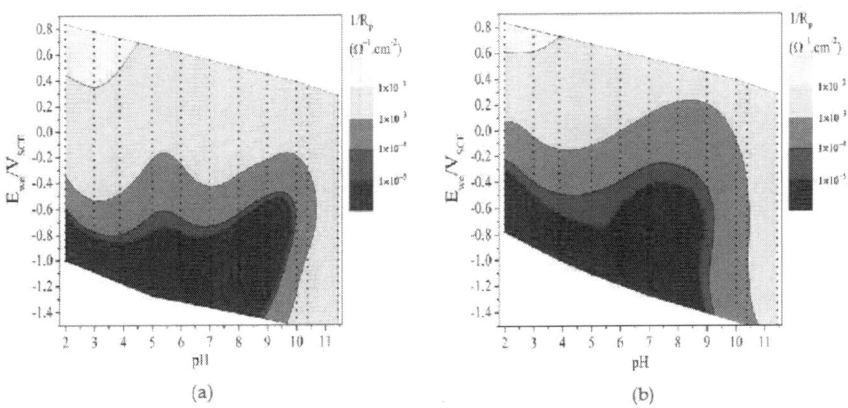

Figure 4. Contour plots of 1/RP for the E-pH space and their data for (a) A portion of a 99.9999% Al ingot single crystal and (b) a polycrystalline specimen from the identical ingot in (a) (adapted from (Zhou, Birbilis et al. 2010)

2.3. The Property Space And Corrosion Property Profile Of Aluminium Alloys

As technologies continue to advance with more challenging applications and environments, a general understanding of durability limits across the class of Al-alloys is essential. Durability needed in a broad sense is the ability to withstand an environment while maintaining mechanical integrity. This indicates a requirement to understand the property space for Al and its alloys. Figure 5 shows a trend that is in line with the notion that increases in hardness (used here as a proxy to yield strength) show increases in corrosion rate.

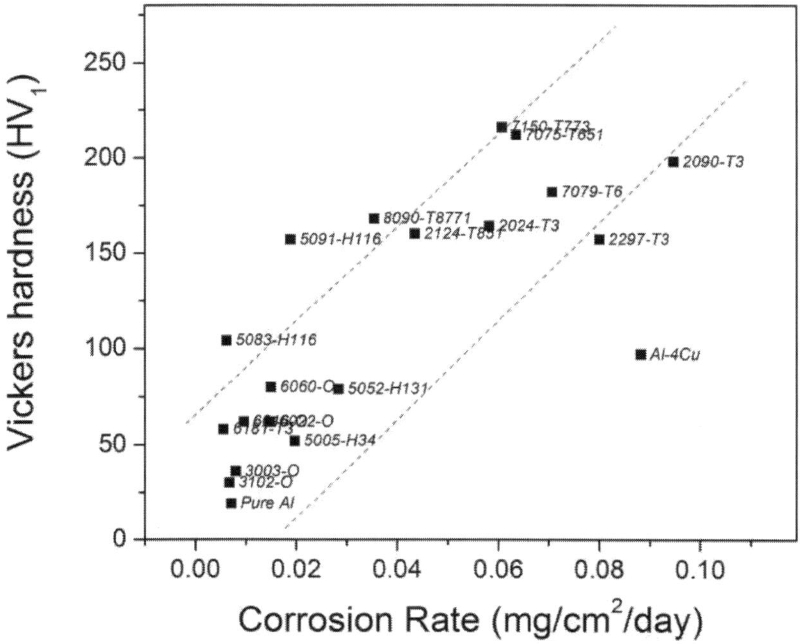

Figure 5. Corrosion rate as determined from weight loss data for commercial Al alloys collected after 14 days exposure in quiescent 0.1M NaCl presented against alloy hardness.

From Figure 5, it can be argued that the data is divided into two main groups, one at each end of the corrosion rate spectrum. High hardness/strength aluminium alloys were found to populate the high corrosion rate space. This is in contrast to the medium to low hardness/strength alloys that revealed considerably lower corrosion rates. The alloys that show the highest corrosion rates are the 'precipitation hardenable' family. Besides the high number density of precipitate particles in such alloys, they also contain an appreciable population of constituent type particles (Chen, Gao et al. 1996;Wei 2001; Andreatta, Terryn et al. 2004; Ilevbare, Schneider et al. 2004; Birbilis, Cavanaugh et al. 2006; Boag, Hughes et al. 2009; Hughes, MacRae et al. 2010; Xu, Birbilis et al. 2011). These particles are industrially necessary, since the complex chemistry of precipitation hardenable alloys (that can contain up to 10 alloying elements) have significant alloying additions added via alloy rich master alloys. It is also observed that the alloys that show the highest corrosion rates also contain appreciable amount of copper.

The plot in Figure 5 allows one to identify a region of property space (at the top left corner) that has potential for future alloys, with ongoing efforts aiming to reach such space (in addition to controlling ductility). Efforts that regard in are underway, focusing on corrosion rate (Carroll, Gouma et al. 2000;Norova, Ganiev et al. 2003; Rosalbino, Angelini et al. 2003; Cavanaugh, Birbilis et al. 2007; Lucente and Scully 2007; Fang, Chen et al. 2009; Graver, Pedersen et al. 2009; Ralston, Birbilis et al. 2010;Tan and Allen 2010; Xu, Birbilis et al. 2011; Brunner, Birbilis et al. 2012) and strength (Poole, Seter et al. 2000; Pedersen and Arnberg 2001; Fuller, Krause et al. 2002; Raviprasad, Hutchinson et al. 2003; Lee, Shin et al. 2004; Oliveira Jr, de Barros et al. 2004; Zhao, Liao et al. 2004; Kim, Kim et al. 2005; Teixeira, Bourgeois et al. 2007; McKenzie and Lapovok 2010; Wang, Zhang et al. 2010; Puga, Costa et al. 2011; Zhong, Feng et al. 2011; Westermann, Hopperstad et al. 2012). However, such studies are done independently of both properties thus the symbiotic effect can't be readily evaluated to date.

3. CORROSION OF ALUMINIUM AND ITS ALLOYS IN AQUEOUS ENVIRONMENT

3.1. The Role Of Chemistry On Corrosion

Alloying elements are added to aluminium for various reasons, with improving mechanical properties the principal reason. These elements introduce heterogeneity into the microstructure, which is the main cause of localised corrosion that initiates in the form of pitting. Each alloying element has a different effect on the corrosion of Al, and in this section we *briefly* discuss the role of alloying elements on corrosion of Al.

3.1.1. Influence of Magnesium

Mg is one of the major elements added to Al to improve mechanical properties by solid solution strengthening – and can be found in 5xxx alloys, as well as 2xxx, 3xxx, 6xxx and 7xxx commercial alloys. Mg can stabilize GP zones, has a high solubility in Al and decreases the alloy density. Muller and Galvele showed that Mg when present in solid solution does not have a significant effect on the pitting corrosion of Al which can be understood on the basis of standard potentials of Al and Mg (Muller and Galvele 1977). Moreover, Mg decreases the rate of the cathodic reaction when present in solid solution, increasing corrosion resistance, which may appear counterintuitive, but is rather obvious (as Mg has a very low exchange current density and hence retards the cathodic reaction). In contrast, excess amounts of Mg in the alloy or a long term exposure to elevated temperature will cause the precipitation of either Al_8Mg_2 or Al_3Mg_2 (Searles, Gouma et al. 2002; Davenport, Yuan et al. 2006;Oguocha, Adigun et al. 2008; Jain, Lim et al. 2012). These phases form typically along grain boundaries (Baer, Windisch et al. 2000; Goswami, Spanos et al. 2010) and are known to be anodic with respect to Al matrix therefore prone to localized corrosion (Vetrano, Williford et al. 1997;Aballe, Bethencourt et al. 2001; Jones, Baer et al. 2001; Brunner, May et al. 2010). Mg in 2xxx, 6xxx and 7xxx alloys however, forms precipitates with other alloying elements to strengthen the alloy where role of Mg mainly depends upon the other alloying additions (Ringer, Hono et al. 1996; Buchheit, Grant et al. 1997; Campestrini, van Westing et al. 2000; Guillaumin and Mankowski 2000;Eckermann, Suter et al. 2008).

3.1.2. Influence Of Silicon

The addition of Si in conjunction with Mg, which is typical in 6xxx series Al alloys, allows Mg_2Si particles to precipitate. There is vast literature on the chemical composition of the Mg-Si phase and its role on mechanical properties (Hirth, Marshall et al. 2001; Usta, Glicksman et al. 2004; Stelling, Irretier et al. 2006; Eckermann, Suter et al. 2008; Zeng, Wei et al. 2011). This particle is beneficial in terms of increasing strength but renders the alloy prone to localised corrosion (Eckermann, Suter et al. 2008). The electrochemical behavior of Mg_2Si was investigated recently and it was shown that Mg_2Si remains more 'anodic' (i.e.. less noble) than the matrix in Al-alloys. As a consequence of this, Mg_2Si remains anodic and undergoes selective dissolution in the Al-matrix. Some 6xxx series alloys contain excess Si. Excess amount of Si however increases the cathodic reaction rate (Eckermann, Suter et al. 2008) and are unfavorable since Si tends to be present along the grain boundary and this may lead to intergranular corrosion and stress corrosion cracking (Guillaumin and Mankowski 2000; Larsen, Walmsley et al. 2008; Zeng, Wei et al. 2011).

3.1.3. Influence Of Copper

The presence of Cu is viewed as detrimental to corrosion due to the formation of cathodic particles capable of sustaining the cathodic reaction locally and efficiently, such as Al_2Cu and $AlCu_2Mg$. In some cases where low Cu content is used, the impact of Cu is minimal, however given that corrosion is not the principal alloy design criteria in most instance, Cu is common in many (most) Al-alloys. The 2xxx series alloys are Cu rich, however Cu is added to other alloy classes such as the 6xxx series where it can increase strength when present in trace amounts, and also enhance precipitation hardening. The same is true in 7xxx alloys, with most modern aerospace alloys having appreciable amounts of Cu that can increase strength by modifying precipitation and minimising SCC via incorporation into precipitates (such as $Mg(Zn,Cu)_2$).

In general however, there is still some debate on the precise role of Cu, which also depends on the temper condition. Muller and Buchheit found that Cu in the form of solid solution decreases pitting susceptibility through the ennoblement of pitting potential. While Muller and Galvele reported an increase in pitting potential for solid solution content of Cu up to 5 wt%. In the case of Al-Cu-Mg alloys which contain *S* phase (Al_2CuMg), large differences in solution potential between Cu (highly noble) and Mg exist, with significant focus on corrosion of *S* phase (Buchheit, Grant et al. 1997;Guillaumin and Mankowski 1998; Buchheit, Montes et al. 1999; Ilevbare, Schneider et al. 2004; Boag,Hughes et al. 2011) revealing dealloying and selective dissolution that leads to preferential dissolution of Mg and Al with Cu remnant being redistributed at or near the site of the Al_2CuMg. A range of other particles associated with Cu have been reported such as Al_7Cu_2Fe. However recent microprobe studies of a number of batches of AA2024-T351 indicate five common compositions across modern alloys which do not have the same composition as older alloy stock indicating that this is still an active area of research (Hughes, Glenn et al. 2012). In general, Cu, or Cu containing particles are capable of supporting high oxygen reductions rates and hence undesirable from corrosion perspective (Mazurkiewicz and Piotrowski 1983; Scully, Knight et al. 1993; Buchheit 2000; Birbilis, Cavanaugh et al. 2006).

3.1.4. Influence Of Zinc

In high strength commercial aluminium alloys such 7xxx series alloys, Zn is added to stimulate precipitation hardening. Alloys containing high levels of Zn such as the modern aerospace alloys 7050 and 7150 are amongst the highest strengths of Al-alloys owing to the high number density of precipitates such as $MgZn_2$ which is evenly distributed throughout the Al matrix (Ringer, Hono et al. 1996; Andreatta, Terryn et al. 2004; Sha and Cerezo 2004; Birbilis and Buchheit 2005; Polmear 2006) in 5xxx alloys. The addition of Zn to Al-Mg alloys was reported to improve resistance against SCC (Unocic, Kobe et al. 2006) where a small amount of Zn added into AA5083 alloy was found to reduce the corrosion - reporting that Zn can promote the formation of Al-Mg-Zn (τ phase) instead of Al_3Mg_2(β phase) the latter of which is responsible for stress corrosion cracking (Carroll, Gouma et al. 2000; Carroll, Gouma et al. 2001).

3.1.5. Influence Of Iron

Iron is typically present as an impurity in all commercial Al alloys due to the production process of Al alloys. Unless specifically required for specialist applications, it is simply too expensive to remove all iron (even in aluminium destined for aerospace applications). Despite having a small fraction of the composition, iron is detrimental to corrosion due to its low solubility and hence ability to form constituent particles which are cathodic to the Al-matrix such as Al_3Fe (Nisancioglu 1990). Additionally, iron is capable of sustaining cathodic reactions more efficiently than Al (Galvele 1976;Szklarska-Smialowska 1999). In more complex alloys, Fe can also combine with other alloying elements such as Mn or Cu (in the latter case forming Al_7Cu_2Fe), which are also a major issue for corrosion (Birbilis, Cavanaugh et al. 2006) since the combination of Fe and Cu provides even greater cathodic efficiency for such particles. Corrosion associated with such noble cathodic constituents/intermetallics leads to an increase in local pH of the solution further enhancing anodic dissolution of the Al matrix adjacent to say, Al_3Fe (Seri 1994; Park, Paik et al. 1999; Birbilis and Buchheit 2005; Ambat, Davenport et al. 2006).

3.1.6. Influence Of Manganese

The addition of Mn is effective in reducing the pitting susceptibility of Al alloys particularly in the context of modifying Fe containing intermetallic particles (Nisancioglu 1990) (where Mn can substitute for Fe, rendering the resulting constituent particle somewhat less noble) The presence of Mn has been noted as reducing the concentration of Fe and reducing the degree of resultant corrosion (Koroleva, Thompson et al. 1999); owing to the formation of Al_6MnFe has a similar electrochemical potential with that of the Al matrix. However, it has also been noted that an excess of Mn can lead to an increase in the cathodic activity when beyond the solubility limit (solubility of Mn in Al is 1.25 wt%) – with constituents such as Al_6Mn readily forming (Liu and Cheng 2010). Generally however, the presence of Mn constituent particles are not as detrimental as particles rich in Fe or Cu (Birbilis and Buchheit 2005; Cavanaugh, Birbilis et al. 2007), which is evidenced by the reliable corrosion performance of 3xxx commercial Al alloy (Zamin 1981; Seri and Tagashira 1986; Tahani, Chaieb et al. 2003; Liu and Cheng 2011).

3.1.7. Influence Of Lithium

The addition of Li in Al alloys is efficient at significantly reducing alloy density whilst increasing strength – making it an obvious contender in the range of transport, namely aero, applications. Al-Li alloys are a rather specialised field that spans the past five decades, with descriptions originally in the 8xxx series compositional space (with principally Li rich compositions). Such so-called 1st generation Al-Li alloys were only used in specialised applications owing to their susceptibility to cracking. The cracking issue was later managed via new alloy compositions and thermomechanical processing (2ndgeneration Al-Li alloys), however until relatively recently Al-Li alloys were not so popular owing to relatively high corrosion rates and localised forms of corrosion propagation. Most recently, the 3rdgeneration of Al-Li alloys has gained significant attention and growing usage in commodity aerospace applications. These 3rd generation alloys are actually 2xxx series alloys, with less Li than Cu. These new 2xxx series alloys will be a significant alloy of the future, whilst still further research is required (from a corrosion perspective) to fully understand the performance, particularly as a function of thermomechanical treatment. Some abridged information regarding the role of Li upon corrosion is included here. In Li rich alloys, the formation of strengthening phase, Al_3Li which is dispersed homogeneously throughout the matrix, is responsible for the increase in strength (Lavernia and Grant 1987). It is however detrimental to corrosion as Al_3Li initially form along the grain boundaries. As Li is an active (i.e. less noble) element, this will localise dissolution to Li rich regions and therefore susceptibility to attack such as intergranular corrosion is high (Martin 1988). When Cu is also added in conjunction with Li (in alloys such as AA2090) the precipitation of phases such as *T1* (Al_2CuLi) occurs. There are two modes of attack associated with *T1*, one of which *T1* at the precipitate free zone is dissolved forming small pits, while the other is when *T1* undergo selective dissolution along with dissolution of the adjacent Al matrix leaving larger pits (Buis and Schijve 1992; Buchheit, Wall et al. 1995).

3.1.8. Influence Of Activating Elements (I.E. Pb, Sn)

Lead (Pb) and tin (Sn) are usually present in low levels as trace elements (Gundersen, Aytaç et al. 2004; Premendra, Terryn et al. 2009). When present in trace amounts, their influence is minimal or negligible. When (by say, recycling or contamination) the levels rise above the solubility limits, the presence of Pb leads to segregation that results in Pb-rich film at the metal - oxide interface when the alloy is heat treated at elevated temperature (Sævik, Yu et al. 2005) causing the Al oxide film to destabilise particularly in the presence of chloride. The disruption of Al oxide film leads to an increase in anodic reaction rate which not only increases the pitting susceptibility, but can activate the entire surface. This process is called anodic activation, and has been well documented for a number of years by studies from the group of Nisancioglu (Keuong, Nordlien et al. 2003; Gundersen, Aytaç et al. 2004;Yu, Saevlk et al. 2004; Yu, Sævik et al. 2005; Walmsley, Sævik et al. 2007; Jia, Graver et al. 2008; Graver, Pedersen et al. 2009; Anawati, Graver et al. 2010; Graver, van Helvoort et al. 2010; Anawati, Diplas et al. 2011). There have been some efforts to reduce the activation effect of Pb by addition of more noble alloying elements such as Cu in the hope that the addition of Cu may alter the surface potential - hence reducing the activation (Anawati, Diplas et al. 2011). A similar result was observed for the addition of Mg (Jia, Graver et al. 2008), however, such methods are not viable on the basis that the Pb interfering with the oxide is an effect in addition to any changes in the alloy potential. The presence of Sn along with Pb however reduces the dissolution rate when annealed at the maximum temperature of 600°C for an hour at which Sn is found to dissolve in the aluminium solid solution diluting the Sn concentration on the surface (Graver, Pedersen et al. 2009).

3.1.9. Influence Of Other Element, Including Zr, Cr, Sc, Ti, W And Sr

These elements are typically added independently in small amounts (i.e. <0.2 wt%) for the purpose of grain refinement, to reduce recrystallisation as well as minimising the effect of intermetallic compounds (Vetrano, Henager Jr et al. 1998). Elements such as Zr and Ti are able to form intermetallics at high temperatures in the Al melt, and persist as finely dispersed particles of Al_3Zr and Al_3Ti within the solidified matrix, which, based on their fine size (i.e. <<1 μm), have a minimal impact on corrosion (Scully, Knight et al. 1993). Similarly with scandium (Sc) additions above the solubility limit the formation of Al_3Sc will occur, contributing to the strength and significantly reducing recrystallisation during thermomechanical processing (Cavanaugh, Birbilis et al. 2007). In general, and neglecting Al_3Fe, such dispersoids are based on the Al_3X system where X is Zr, Ti, Sc, Er, etc, and taking the form of fine insoluble dispersoids which are functional in grain inoculation and refinement. As such, there are specific alloying additions of Ti and Zr to high strength alloys such as AA7075 (Senkov, Bhat et al. 2005; Zou, Pan et al. 2007; He, Zhang et al. 2010).

Whilst not studied in detail, it has been posited that the ability to suppress recrystallisation leads to lower corrosion rates by avoiding the formation of high angle grain boundaries (Fang, Chen et al. 2009). Furthermore, there are also complex second and third order interactions between sparingly soluble elements that extend beyond the predictions of simple phase diagrams. An example is that the addition of Sr will impact intermetallics such as Al_5FeSi, making them smaller in both their size and volume fraction (Ashtari, Tezuka et al. 2003; Eidhed 2008). Such an effect has a role in corrosion by minimising the number of intermetallic sites. The purpose however of this section, is not to describe the metallurgy, as that has been done classically as far back as Mondolfo (Mondolfo 1971; Mondolfo and Barlock 1975), but to emphasise the microstructures direct impact on corrosion.

In common alloys prepared by conventional casting technologies, transition metals (TMs) such as W, Mo and Cr are not employed owing to their very low solubility limits. However it is important to note that when prepared in sputtered or thin film forms, such Al-TM alloys display the lowest corrosion rates of all the Al-alloys. Shaw successfully produced the alloys by sputter deposition and found that these elements increased the pitting potential and passivity of the alloys as well as inability to form second phase upon heat treatment (Shaw, Fritz et al. 1990; Shaw, Davis et al. 1991). The work of Frankel also showed promising results (Frankel, Russak et al. 1989; Frankel, Davenport et al. 1992;Tailleart, Gauthier et al. 2009).

3.2. The Role Of Microstructure On Corrosion

In order to understand the corrosion performance of Al alloys, and following on from the previous section, an appreciation of the microstructure is vital. Alloying elements and thermomechanical processing play an important role in dictating the type of microstructure produced. For homogeneous alloys, such as pure Al or 5xxx series alloys, corrosion susceptibility is low due to lack of pre-existing microstructural attack sites. The main concern however is regarding heterogeneous alloys, particularly the higher strength Al alloys such as the 2xxx, 7xxx and heat-treatable 6xxx series, where microstructural heterogeneity is a necessity. The most common features of a microstructure are the intermetallic particles which are classified into precipitates (forming from nucleation and growth, nominally 1nm to 300nm in diameter), constituent particles (from insoluble or impurity elements, unable to redissolve, nominally a few microns, to a few tens of microns, in size) and dispersoids (nominally $\ll$ 1 micron in size) (Polmear 2006). Each of these features consists of different electrochemical characteristics (including their native electrochemical potential and the currents they can sustain at a given potential characteristic of the alloy which they populate) and act as the sites which dictate the severity of corrosion attack. Work on categorising such intermetallics in relation to corrosion is plentiful and now has been going on for several decades (Mazurkiewicz and Piotrowski 1983; Scully, Knight et al. 1993; Buchheit, Grant et al. 1997; Birbilis and Buchheit 2005; Eckermann, Suter et al. 2008; Goswami, Spanos et al.

2010; Boag, Hughes et al. 2011; Hughes, Boag et al. 2011). We make the distinction in this chapter that whilst such intermetallics are responsible for corrosion initiation steps, including pitting, we will not cover corrosion propagation in detail (i.e. stress corrosion, or intergranular corrosion) since they would require a dedicated chapter.

The knowledge of intermetallic chemistry and the electrochemistry allows a prediction of the mode of corrosion and the propensity of the attack (Cavanaugh, Buchheit et al. 2009). Theoretically, a more active particle (i.e. less noble) will become a local anode and consequently corrode while the more noble particles become cathodes (Szklarska-Smialowska 1999). This is not always a true reflection of kinetics however, as reported by Birbilis where the capability of the element to sustain the cathodic reduction process can not be the deduced by relative nobility as well (Birbilis and Buchheit 2005). As corrosion occurs upon Al alloys, particularly pitting and early damage accumulation, two types of corrosion mode are identified. In one mode of attack whereis the intermetallic is classified as a cathode, the surrounding matrix tends to corrode leaving a ring shape around the particle or also called trenching. There is still some uncertainty on whether or not the trench itself is a result of microgalvanic coupling alone, or if the major contributor is local pH elevation, however a good treatise of this topic was given in a multi-part series of papers by the group at Virginia (Ilevbare,Schneider et al. 2004; Schneider, Ilevbare et al. 2004; Schneider, Ilevbare et al. 2007). In some cases damage may propagate to the base of the particle and eventually lead to particle fall out (Buchheit, Grant et al. 1997). The other mode of attack is when the intermetallic acts as anode and matrix as cathode; whereby the intermetallic will corrode leaving a cavity on the surface. Finally, in some microstructurally complex systems with ternary and above alloying additions, another type of attack found to not follow the traditional way of determining anode and cathode is incongruent dissolution commonly found in 2xxx and 7xxx due to the presence of *S* phase (Al_2CuMg) (Buchheit, Grant et al. 1997; Guillaumin and Mankowski 1998). This intermetallic contains Cu and Mg, whereby (Blanc, Lavelle et al. 1997; Buchheit, Martinez et al. 2000)., *S* phase experiences selective dissolution of the highly active Mg hence leaving Cu remnants at the bottom of the pit (Buchheit, Grant et al. 1997;Büchler, Watari et al. 2000) or redistrubuted near the particle site. Although the more detailed study of Boag et al. indicated that Al might be preferentially removed in the initial stages of attack on S-phase (first 2.5 minutes) as both Cu and Mg were observed to be present after Al dissolution at 5 minutes. This may have been due to a combination of the types of aluminium hydroxyl-chloro complexes that form and the partial switching of the areas of S-phase to Cu remnants where cathodic reactions lead to the formation of insoluble $Mg(OH)_2$ (Boag, Hughes et al. 2011). After 15 minutes the Mg is removed as well and no chloride was detected on these particles. The attack then continues with the dissolution of Al matrix (Buchheit, Grant et al.

1997; Guillaumin and Mankowski 1998). Localized attack of the intermetallic also influenced by the chloride content and pH of the electrolyte. Higher chloride content is widely reported to have higher pitting occurrence due to passive layer disruption of chloride ions (Seri 1994; Blanc, Lavelle et al. 1997; Ilevbare, Schneider et al. 2004) but this does not necessarily mean the pit will propagate deeper (Cavanaugh, Buchheit et al. 2010). There exists a dedicated monograph on this topic (Muster, Hughes et al. 2009). It has also been noted that the intermetallic Mg_2Si can undergo similar incongruent dissolution, whereby Si enrichment occurs at the expense of dissolving Mg (Birbilis and Buchheit 2005; Jain, 2006; Eckermann, Suter et al. 2008; Gupta, Sukiman et al. 2012).

The revelation of a large number of microstructure vs. corrosion micrographs will not be done herein, however if the readers are interested, a nice expose of such images exists in Cavanaugh (Cavanaugh 2009). Instead, a demonstration is given here. Figure 6 shows the micron scale microstructure for 2024-T3 and 5083-H116 before and after corrosion exposure in 0.1M NaCl for 14 days. These relatively low magnification images do not reveal the precipitate structure in AA2024-T3, instead showing the coarse intermetallics that exist in the alloys. What is observed is that a higher intermetallic density gives more possible sites for localised attack. In the relatively widely studied 2024-T3 (Guillaumin and Mankowski 1998; Schmutz and Frankel 1998; Campestrini, van Westing et al. 2000;Leblanc and Frankel 2002; Boag, Hughes et al. 2009; Hughes, MacRae et al. 2010; Ralston, Birbilis et al. 2010; Boag, Hughes et al. 2011; Hughes, Boag et al. 2011; Zhou, Luo et al. 2012). pitting attack is associated with the Cu containing intermetallic such as *S* phase (coarse and precipitated) and the intermetallic density in AA2024-T3 is high, owing to the very large number of alloying elements which can leave the possibility of forming constituent particles based on Fe, or Si, or Cu, or Mn, etc. In contrast to AA2024, the alloy 5083-H116 with a smaller number of alloying elements, and Cu free, has a lower constituent number density. in addition to the Mg remaining in solid solution and reduces the susceptibility to localised attack (Vetrano, Williford et al. 1997; Aballe, Bethencourt et al. 2001;Yasakau, Zheludkevich et al. 2007).

4. CORROSION PROTECTION

This section covers general approaches to protection of aluminium alloys in view of recent advances in the understanding of alloy microstructure. It includes an overview of pretreatment processes such as anodising, conversion coating and organic coatings (barrier and inhibitor combinations). It will examine recent advances in inhibitor design such as building in multifunctionality and touch upon self-healing coating systems.

Approaches using multifunctionality can target anodic and cathodic reactions more effectively than using individual monofunctional inhibitors.

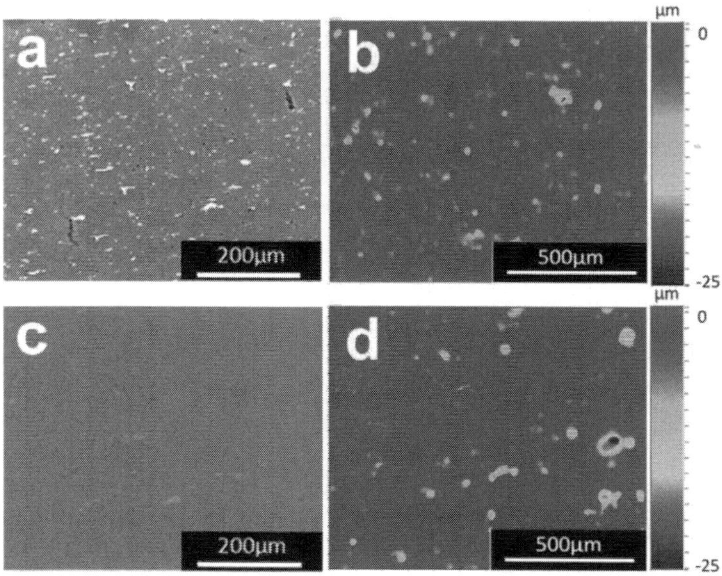

Figure 6. SEM images for AA2024-T3 and AA5083-H116 before exposure in (a) and (c) and after exposure in 0.1M NaCl for 14 days in (b) and (d). These are comparative images to show the extent of damage following immersion.

Standard metal finishing processes, which have been used for many years, are likely to continue to be used into the future unless they contain chemicals that are targeted for replacement such as chromium. The function of these coatings is primarily to provide better adhesion properties for paint coatings and a secondary role is to provide corrosion protection. The general approach for applying these coatings relies on metal finishing treatments (treatment prior to painting involving immersion in acidic and alkaline baths)) with the objective of reducing the heterogeneous nature of the metal surface such as removing the NSDL and second phase particles (Muster 2009). This is achieved in multistep treatment processes for metal protection (Twite and Bierwagen 1998; Buchheit 2003; Muster 2009) as for instance:

- selective deoxidation (IM particle removal and surface etching);
- deposition or growth of a manufactured oxide via electrochemical (anodising) or chemical (conversion coating) means;

- use of an organic coating for specific applications, normally including a primer and a top-coat.

On aluminium, most anodised coating processes produce an outer oxide with a cellular structure on top of a thin barrier layer that provides some protection against corrosion. Inhibitors can be incorporated into the outer porous layer of the anodized layer during formation or as a seal after formation to offer some extra protection upon damage. Chromic acid anodizing is one of a number of processes that are available for electrochemical growth of surface protective oxides. More environmentally friendly alternatives to chromic acid anodizing such as sulfuric, sulfuric-boric, sulphuric-tataric and phosphoric based processes have been available for a long time. There have been a number of recent advances in reducing the energy consumption of anodizing processes as well as improving coating properties. These advances are based on an improved understanding of the alloy microstructure described above and involve selective removal of second phase particles as part of the anodising process.

An alternative approach to anodizing is to precipitate a coating on a surface through chemical means called conversion coatings. For high strength Al-alloys such as 2xxx and 7xxx series chromate conversion coating (CrCC) is still the preferred process. Replacements for the toxic chromate-based conversion coatings include a range of treatments based on self-assembled monolayers, sol-gel chemistries, Ti/Z oxyfluorides, rare earth, cobalt, vanadates, molybdates and permanganate processes (Twite and Bierwagen 1998; Buchheit 2003; Kendig and Buchheit 2003). These processes are widely developed for chemically pretreated surfaces that have nearly all the IM particles removed (by chemical pretreatment) and are not specifically designed to address electrochemical and compositional variations found for a heterogeneous surface such as when the IM phases are present. Work like that in (Buchheit and Birbilis 2010) depicting the reaction rate variation across the surface, however, opens an avenue to start designing inhibitors where the initial reaction rate distribution across a surface can be significantly reduced to limit the overall activity of the surface. In this context reaction of inhibitive phases with manufactured IM compounds as well as IM particles within the alloy have been studied for a number of systems (Juffs, Hughes et al. 2001; Juffs 2002; Juffs, Hughes et al. 2002; Birbilis, Buchheit et al. 2005; Scholes, Hughes et al. 2009).

Once the anodised or conversion coating is applied, the surface is ready to receive the organic coating. There are many different types of organic coatings, however because of the focus on 2xxx and 7xxx alloy used in the aerospace industry this section will only deal with that application area. The organic coating system usually consists of a primer and a topcoat. The primer is the main protective layer including corrosion inhibitors that can be released when corrosive species or water reach the metal. From the perspective of providing protection for the underlying aluminium alloy, the inhibitor needs to be available during a corrosion event at a concentration higher than the minimum concentration at which the inhibitor stops corrosion (critical concentration). While this sounds obvious, the critical inhibitor concentration needs to be maintained over many years for structures such as airframes, where maintenance may not be possible in parts of the aircraft because of poor access. The chromate systems itself provide continuous protection and repair to the surface for as long as the dose of chromate remains above the critical concentration. This mechanism of inhibitor release and metal protection is recognized as a self-healing mechanism, since the release of the active species recovers the protective layer on top of the metal.

The search for green inhibitors as replacements for chromate has been driven by legislative imperatives for a number of years. Needless to say, replacement inhibitors do not have the same intrinsic inhibitive power at low solubility as chromate. Thus solubility, inhibitive power and transport within the primer system (which consists of a number of inorganic phases as well as the epoxy) ultimately mean that finding a replacement for chromate is difficult. This means that alternatives must be present at higher concentration leading to the use of more soluble compounds and consequently encapsulation as a method of regulating the response to external or internal triggers emerges as a prospective way to achieve this objective.

Many current inhibitors are water soluble salts and thus ionic. Consequently, they exist as either anions or cation in solution and perform the single function of anodic or cathodic inhibition. So the simplest improvement to inhibitor design is to increase the functionality by finding compounds which play both a cationic and anionic inhibitive role. A large range of cations including Zn, Ca, and rare earths (Bohm, McMurray et al. 2001; Du, Damron et al. 2001; Kendig and Buchheit 2003; Taylor and Chambers 2008; Muster, Hughes et al. 2009) have been combined with either organic (Osborne, Blohowiak et al. 2001; Sinko 2001; Voevodin, Balbyshev et al. 2003; Khramov, Voevodin et al. 2004;Blin, Koutsoukos et al. 2007; Taylor and Chambers 2008; Muster, Hughes et al. 2009) or inorganic (oxyanions, carbonates, phosphates, phosphites, nitrates, nitrites, silicate (Bohm, McMurray et al. 2001; Sinko 2001; Blin, Koutsoukos et al. 2007; Taylor and Chambers 2008) compounds.

Anions with dual functionality, such as some of the transition metal oxyanions which are both oxidants and anions, have been investigated extensively. The oxidizing oyxanions or some organophosphates have some degree of bio-inhibition required for some applications. Substitution of different organophosphates into rare earth-based inhibitors provide versatility in designing inhibitors for specific applications (Birbilis, Buchheit et al. 2005; Hinton, Dubrule et al. 2006; Ho, Brack et al. 2006; Blin, Koutsoukos et al. 2007; Markley, Forsyth et al. 2007; Markley, Hughes et al.

2007; Forsyth, Markley et al. 2008; Deacon, Forsyth et al. 2009; Scholes, Hughes et al. 2009). Thus Ce(di-butyl phosphate)$_3$ is a good inhibitor and relatively "green" whereas Ce(di-phenyl phosphate)$_3$ is also a good inhibitor, but the diphenyl phosphate also has strong bio-inhibition characteristics (García 2011). However, good bio-inhibition usually means that there are increased environmental and health risks. Obviously the number of cathodic and anodic inhibitors means that there are an enormous number of possible combinations, particularly if ternary and quaternary combinations are considered. Hence high-throughput techniques are being used to assess new inhibitor.

As pointed out above, the kinetics of inhibitor release are of the utmost importance since the inhibitor should be available at levels above the critical inhibitor concentration. Optimization of the release kinetics by novel delivery systems becomes integral to incorporation of new inhibitors.

There are a number of different mechanisms investigated for release of healing agents or corrosion inhibitors which can be incorporated into organic coatings. Both mechanical damage and water are triggers for inhibitor release. In the former case mechanical damage breaks capsules containing water soluble inhibitors. In the latter case water dissolves inhibitor directly incorporated in the primer. Droplet formation within defects such as scratches means that the inhibitor is only released when required i.e., when the defect is moist (Furman, Scholes et al. 2006). There is some evidence to suggest that initial high release of inhibitors may be facilitated through atmospheric exposure of the intact paint where penetration of water into the film "prepares" the inhibitor, probably via surface hydrolysis reactions, within the paint, for diffusion and release into the defect (Joshua Du, Damron et al. 2001; Furman, Scholes et al. 2006; Scholes, Furman et al. 2006; Souto, González-García et al. 2010). The presence of water in the film allows soluble inhibitor species to be released into the paint system and diffuse to the metal/coating interface to provide in-situ corrosion prevention or repair called pre-emptive healing (Zin, Howard et al. 1998; Osborne, Blohowiak et al. 2001; Mardel, Garcia et al. 2011). Thus it has been demonstrated that water can trigger cerium dibuthylphosphate (Ce(dbp)$_3$) release into an epoxy matrix resulting in improved adhesion and resistance to filiform corrosion attack through interfacial modification (Mardel, Garcia et al. 2011)

In terms of delivery systems, hard capsules, which have been used in polymer healing (Dry 1996;White, Sottos et al. 2001; Mookhoek, Mayo et al. 2010) need to be smaller for paint systems particularly in the aerospace industry where coatings are typically 20 μm or less (Yin, Rong et al. 2007; Fischer 2010; Hughes, Coles et al. 2010; Mookhoek, Mayo et al. 2010). In polymer applications, capsules up to a few hundred microns can be accommodated (Yin, Rong et al. 2007; Wu, Meure et al. 2008; Tedim, Poznyak et al. 2010). The concept of encapsulation has already been successfully applied to protective organic coatings under different concepts: i) liquids filling completely the void created by the damage by adopting a bi-component systems where one component is encapsulated and the other distributed in the matrix (Cho, White et al. 2009), or single based components with water reactive oils like linseed and tung oils (Suryanarayana, Rao et al. 2008; Samadzadeh, Boura et al. 2010)and ii) liquids (i.e. silyl esters) forming a hydrophobic and highly adhesive layer covering the metallic surface by reaction with the underlying metal and the humidity in air (García, Fischer et al. 2011). One adaption for capsules is to increase the volume of self-healing material by manufacturing rods instead of spheres. Rods with the same cross-sections as spheres can deliver larger volumes of material (Bon, Mookhoek et al. 2007; Mookhoek, Fischer et al. 2009). For inhibitors, their role is to prevent a surface reaction (corrosion) and therefore, the volume of material required is much smaller than that required to actually fill the defect. Consequently, there has been considerable effort looking at "nano-containers" (Voevodin, Balbyshev et al. 2003; Raps, Hack et al. 2009; Tedim, Poznyak et al. 2010).

Water is the most obvious trigger since it can permeate most polymers. pH variations are more specific and respond to the pH excursions that occur in corrosion reactions and by an understanding reactions that occur at different sites in the alloy microstructure. The presence of chloride ions (and other anions) within the coating can be used as specific triggers for the release of corrosion inhibitors and uptake of corrodents using anion exchange materials, such as layered double hydroxides (e.g. hydrotalcites) (Tedim, Poznyak et al. 2010) (Bohm, McMurray et al. 2001; Buchheit, Guan et al. 2003; Williams and McMurray 2003; Zheludkevich, Salvado et al. 2005; Mahajanarn and Buchheit 2008). In this context hydrotalcites have been loaded with vanadate, chromate, nitrate and carbonate which exchange for chloride ions and prevent interfacial damage (Bohm, McMurray et al. 2001;Williams and McMurray 2003; Mahajanarn and Buchheit 2008). The incorporation of Mg particles into paint act as sacrificial anodes to protect Al alloys and steels (Battocchi, Simoes et al. 2006).

5. RECENT ADVANCES IN ASPECTS RELATED TO CORROSION OF ALUMINIUM ALLOYS

The search for new multifunctional inhibitors has led to the development of high throughput and combinatorial assessment of new combination of inhibitors. These include multielectrode techniques, and high throughput versions of standard corrosion tests. A range of new electrochemical techniques including AC/DC/AC, SVET, LEIS, SECM and SIET will also be described.

Since corrosion of aluminium alloys tends to be dominated by electrochemical processes, most of the techniques employed for the evaluation of corrosion and protection are based on electrochemical approaches. Furthermore, combining electrochemical techniques with other microscopic, analytical and spectroscopic techniques enables the identification of corrosion products in solution (such as inductive coupled plasma (ICP) and UV-Vis). This combination provides an even broader mechanistic understanding of the level of corrosion and/or corrosion protection.

The increasing number of corrosion inhibitor alternatives to chromates has boosted interest in developing high-throughput techniques and combinatorial assessment of new corrosion inhibitors in aqueous solution. At the same time, the traditional techniques (accelerated or not) employed in the evaluation of the performance of organic coatings require long evaluation periods and are relatively expensive to run, and only offer qualitative or semi-quantitative information at best (e.g. salt fog spray tests). For these reasons, new accelerated techniques for the evaluation of coating performance that offer quantitative results are needed.

Figure 7 shows a simplified flowchart for the formulation of anticorrosive (organic) coatings. The chart includes some of the most common techniques employed in corrosion inhibitor and coating performance evaluation. In the figure, the parallel and complex line of the development of the polymeric matrix (i.e. organic coating) is not included, but awareness of its existence is important, since factors such as the corrosion inhibitor/coating matrix compatibility should be taken into account. For the development of anticorrosive organic coatings, several steps are proposed:

Formulation of inhibitors:

The number of corrosion inhibitor candidates is virtually unlimited, and is motivated by the urgent need to replace chromate based inhibitors by environmentally friendly and non-toxic ones, as well as the development of new concepts such as self-healing and synergies between anodic and cathodic inhibitors which open up the broad range of possibilities of organic chemistry. One example of the complexity of the introduction of organic compounds as corrosion inhibitors is the effect that the position of certain groups in a cyclic organic compound can have in the corrosion protection efficiency (Harvey, Hardin et al. 2011).

Evaluation in aqueous solution:

Once the inhibitors have been formulated, they can be tested by means of traditional aqueous solution tests such as electrochemical impedance spectroscopy (EIS), potentiodynamic polarisation (PP), immersion tests and weight loss/gain. Also local electrochemical techniques (see point 4- evalaution of organic coatings) give very important information of the mechanisms of corrosion protection offered by the different species in solution. Since traditional techniques require long periods of time and a large number of samples, the introduction of high- throughput techniques as a preceding step is important in order to reduce the number of inhibitors that enter further evaluation processes using traditional aqueous solution tests. It is necessary to highlight that high-throughput techniques are not aimed at replacing traditional tests but at complementing them in order to reduce cost and time in the corrosion inhibitor selection process. Some examples are: single metal, multielectrode array (Chambers, Taylor et al. 2005; Chambers and Taylor 2007), microchannels (White, Hughes et al. 2009) and multi metal multielectrode (Muster, Hughes et al. 2009; García, Muster et al. 2010; Kallip, Bastos et al. 2010).

Introduction into an organic matrix:

The introduction of pigments into organic coatings adds some extra difficulties to the whole process, leading to a lot of extra research to avoid undesirable reactions between the polymer matrix and the inhibitors. Some of the parameters to take into account are the ratio between the pigment volume concentration (PVC) and the critical pigment volume concentration (CPVC), the possible side reactions between the polymer and the pigment, with consequences in parameters like the barrier properties, gloss, active corrosion protection, and adhesion amongst others. At the same time, parameters such as contaminant reduction and parameters related to the polymeric matrix itself (such as adhesion and the glass transition temperature (Tg)) should be considered. In any case, once the pigments have been introduced into the organic coating, the coating's performance has to be tested for protection efficiency and if results are promising, then start the optimization process.

Evaluation of the performance of organic coatings:

As in the case of aqueous solution tests for evaluation of corrosion inhibitors, the introduction of accelerated tests to evaluate coatings performance is necessary to reduce the amount of time and number of samples that move into traditional assessment. Several techniques have been proposed in this direction, such as the technique AC/DC/AC (Hollaender 1997; Bethencourt, Botana et al. 2004;Rodriguez, Gracenea et al. 2004; Garcia and Suay 2006; Garcia and Suay 2006) (García, Rodríguez et al. 2007; Garcia and Suay 2007) (Garcia and Suay 2007; García and Suay 2007; García and Suay 2009), (Poelman, Olivier et al. 2005; Allahar, Bierwagen et al. 2010; Allahar, Wang et al. 2010), and thermal cycling (Bierwagen, He et al. 2000). Also new concepts like the use of flow induced degradation (Wang 2009) are interesting for developing accelerated testing techniques.

Optimisation:

Before entering the pre-commercialization phase, the final step of anticorrosive coatings formulation is the optimization, which is the improvement of the system by modifying pigments concentration, type, and delivery systems to improve and extend the service lifetime protection and compatibility with the matrix. This step is iterative as shown in Figure 7. The whole process from conception to commercialization of the system can take several years.

Due to the impossibility to cover the broad amount of existing (new) high-throughput techniques for selection of corrosion inhibitors for aluminium alloys and accelerated tests for evaluation of protective organic coatings on aluminium alloys, we focus on two electrochemical techniques that have attracted a broad interest most recently due to their high potential and relatively well understood evaluation procedure.

5.1. High Throughput Assessment

In terms of high-throughput techniques, those based on electrochemical approaches are of most interest, since their measurement principles are closely related to familiar research techniques such as Electrochemical Impedance Spectroscopy (EIS) and potentiodynamic polarisation (PP), while at the same time offer quantitative information about the corrosion and corrosion protection mechanisms.

Chambers and Taylor (Chambers, Taylor et al. 2005; Chambers and Taylor 2007) first presented the use of multi-electrode arrays for rapid screening of corrosion inhibitors in different concentrations and pHs, putting identical pairs of AA2024-T3 wires into a large number of separate reaction cells containing different inhibitive solutions and applying a 100mV potential between the two wires in each cell (Figure 8(a)). The current between both electrodes was used to determine the polarisation resistance. With this set-up they were able to evaluate for one metal many different combinations of inhibitors and their synergistic behavior, while at the same time showing its possible application for evaluating the influence of the pH on inhibitors' efficiency.

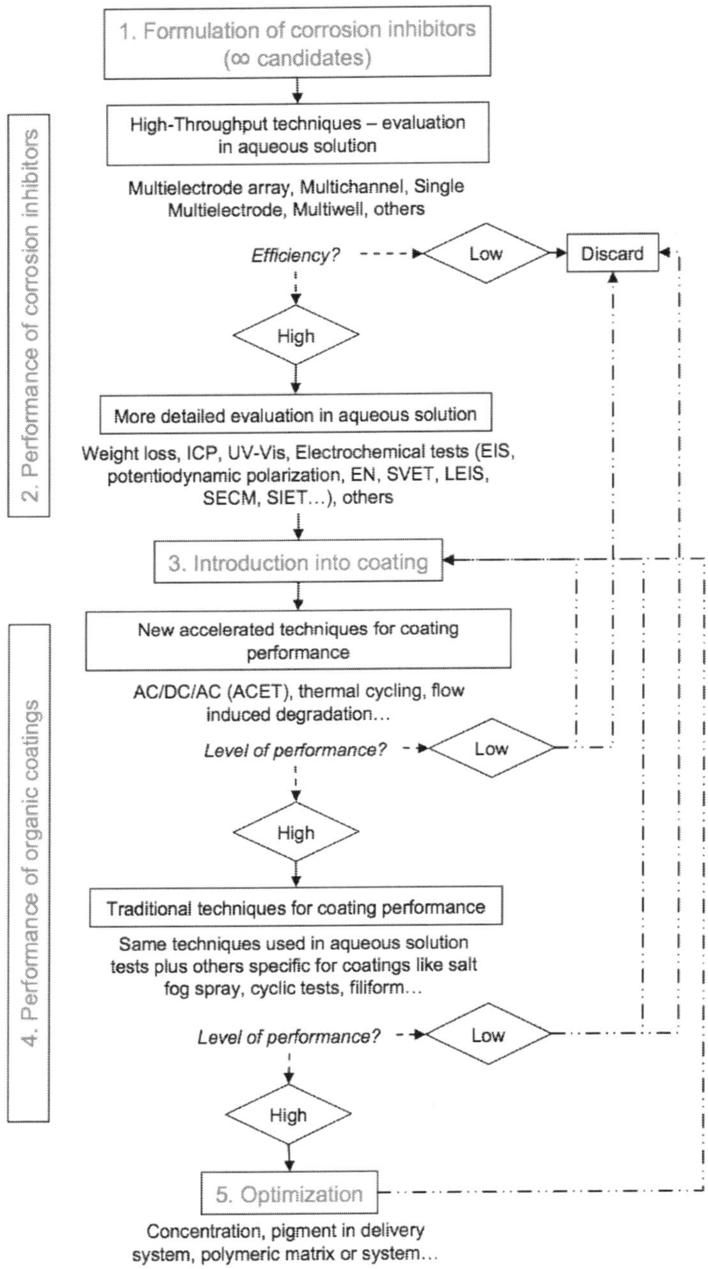

Figure 7. Simplified flowchart for anticorrosive coatings development departing from the inhibitor design or formulation.

Based on the work of Taylor and Chambers, Muster et al. (Muster, Hughes et al. 2009) proposed a variation of the method using a combination of different pairs of metals assembled together (Figure 8(b)) to form what was presented as a multielectrode (ME). The basic idea of this ME was to test a combination of nine pairs (or as many as are interesting) of different metals in the same electrode configuration connected by means of a multiplexer to a potentiostat/galvanostat. Measurements consist of applying 100mV between a selected pair of the same metals within the ME and measure the current flow between them, repeating afterwards the same procedure for the other metal pairs. This set up was employed to rapidly evaluate, without removing the ME from the solution, the concentration range at which a particular inhibitor or combination of inhibitors were offering corrosion protection. Hence, the setup allowed the determination of optimal metal-inhibitor combinations, while significantly reducing the evaluation time with respect to conventional PP tests, without the need of a reference electrode, which simplifies the experimental setup.

In a second paper (Garcia, Muster et al. 2010), the authors studied the effect of the pH and inhibitor type on the correlation between the ME and PP for AA2024-T3. The findings were promising due to the high level of correlation between the ME and traditional techniques, although some discrepancies were found for corrosion inhibitors that can speciate or precipitate at certain pHs. Nevertheless, the non- correlation was assumed to be dependent on the type of corrosion inhibitors and not due to conceptual or experimental mismatching between techniques.

A second concern with the ME was the possibility of cross-contamination due to the presence of several metals in one solution. Garcia et al. (Garcia, Muster et al. 2010) also addressed this problem studying the effect of cross-contamination for the AA2024-T3 wires within the ME. This study showed that if there was cross-contamination then it was not significant enough to influence the results. Despite these results, some more studies should be performed with the ME to check cross-contaminations for other metals such as AA7075-T6 which could be more susceptible to copper plating coming from other metals such as AA2024-T3.

The results obtained so far with multielectrodes (multielectrode array and ME) are very promising and relatively extended information can be found in literature. Nevertheless, more studies and data treatment simplification need to be performed to completely validate these techniques and lead them to an industrial application level.

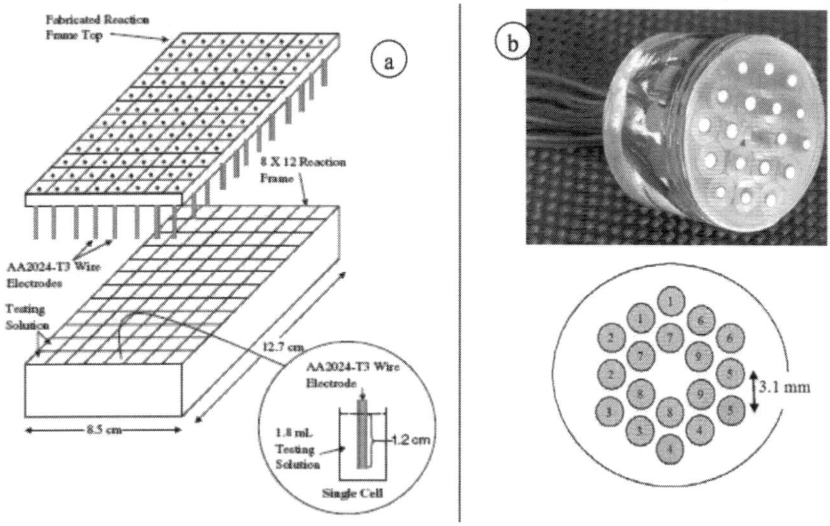

Figure 8. a) single metal multiarray (Chambers and Taylor 2007); b) multimetal multielectrode (Muster, Hughes et al. 2009)

5.2. Ac/Dc/Ac Accelerated Technique For Coating Evaluation

This technique has already reached its maturity and is actually employed at an industrial level under the name of accelerated electrochemical technique (ACET) (Medco). Although the industrial application differs from the research technique (AC/DC/AC), the evaluation and testing are conceptually similar. The early version of the AC/DC/AC technique was performed for the packaging industry (Hollaender 1997). The accelerated technique developed by Hollaender (named AC/DC/AC) was based on the use of temporary stresses to accelerate degradation, and consisted of a first EIS test (AC) to evaluate the initial state of the coating, followed by a cathodic polarisation (DC) and a new EIS (AC) to detect the degradation of the lacquer.

The work initiated by Hollaender was further developed by Suay, Garcia and Rodriguez who successfully applied a modified version of the AC/DC/AC technique to evaluate the performance of organic coatings for carbon steel protection and compared the obtained results with EIS, salt-fog spray, and cyclic tests. The technique was then tested for liquid paints (Bethencourt, Botana et al. 2004;Rodríguez, Gracenea et al. 2004), powder coatings (García and Suay 2006; García and Suay 2006;García and Suay 2007; García and Suay 2007) and cataphoretic paints to optimize parameters such as cataphoretic potential and curing time (Poelman, Olivier et al. 2005; García, Rodríguez et al. 2007;García and Suay 2007; García and Suay

2009). The new version of the technique included a crucial step: the relaxation of potentials (open circuit potential relaxation) after the application of each cathodic polarisation. Furthermore, the potentials applied during the cathodic polarisation depended on the type of coating that was studied, although -4V for 20 minutes was preferred. Figure 9(a) shows a schematic of the AC/DC/AC technique procedure, including the relaxation step proposed by Garcia and Suay. The AC/DC/AC cycle is repeated 6 times (6 cycles) leading to a testing time per sample of around 24 hours, which is a significant improvement when compared to traditional EIS and salt-fog tests which require weeks or months. Nevertheless, depending on the quality of the coatings the number of cycles could be increased and the relaxation time reduced or extended.

The cathodic polarisation aims to degrade the coating and coating-metal interface (e.g. pore formation and delamination) due to hydrogen and OH− production. If the coating is good then it has a higher number of cycles to degrade and the effects in the impedance and potentials relaxation are less pronounced, while a lower quality coating will display a faster degradation. An example of these effects compared to traditional EIS and salt-fog spray is presented by Garcia et al. (García and Suay 2009)

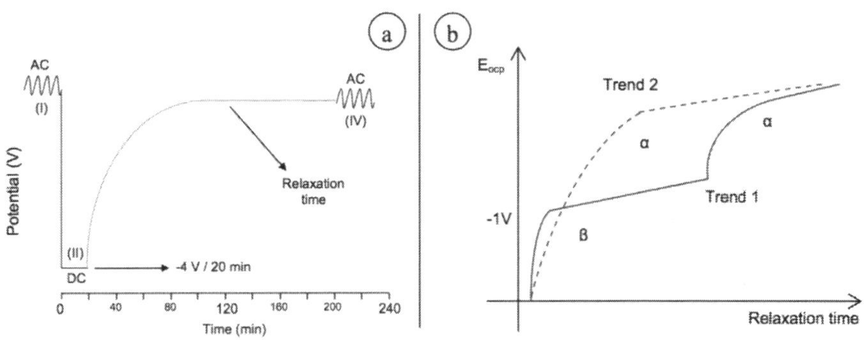

Figure 9. a) AC/DC/AC testing scheme (García and Suay 2009); b) Trends in relaxation of potentials.

Apart from the EIS spectral evolution with cycles, for which a higher drop of impedance is related to higher degradation, the evolution of the open circuit potential after polarisation (relaxation of potentials with time) provides extra crucial information about the degradation of the systems under study (García and Suay 2006). When the cathodic polarisation finishes, the potential of the system relaxes leading to two types of trends depending on the quality of the film (Figure 9(b)):

a. Strongly degraded systems (Trend 1, Figure 9(b)), show two time relaxations of the potential, namely α and β. The first relaxation in time (β) is related to the end of the cathodic reactions that took place at the metallic surface. This relaxation is observed as a quick relaxation around −1V (with small variations depending on the system). The second relaxation (α), which occurs later in time, corresponds to ions and electrolyte leaving the coating. The relaxation (β) could not be detected in all cases due to extremely long relaxation times, although if enough time was allowed this relaxation was then detected, supporting the idea of the two relaxation processes (García and Suay 2009)

b. Less degraded systems (Trend 2, Figure 9(b)), only show relaxation α which is detected at relatively short times of relaxation. This relaxation will take place at longer times as ions and electrolyte penetrate deeper into the film. More recently, Allahar et al. (Allahar, Wang et al. 2009; Allahar, Upadhyay et al. 2010; Wang, Battocchi et al. 2010) have performed an extensive study on the understanding of the AC/DC/AC technique and its principles, leading to a broader understanding of the relaxation of potentials and supporting previous theories, while at the same time further validating the technique.

In this section we have highlighted the complexity of the selection of corrosion inhibitors and anticorrosive coatings formulation leading to the design and validation of high-throughput and accelerated techniques, which at a certain stage can become commercial techniques. The interest in developing new techniques and understanding those already existing is indeed growing in recent years due to the need of developing environmentally friendly and non-toxic systems.

5.3. Staircase Impedance

Staircase potentio-electrochemical impedance spectroscopy (SPEIS) is based on EIS, nowadays common for investigating electrochemical and corrosion system. The basis of EIS is by measuring impedance over a range of frequency and the data obtained is expressed graphically in Bode or Nyquist plot format. As for SPEIS, it is designed for impedance measurements over a range of frequency as well as over a range of voltages (similar to the signal sequence used for Mott-Schottky analysis (Barsoukov and Macdonald 2005; Orazem and Tribollet 2008). The potential range is divided into several potential steps and each step contains a DC relaxation period for a given rest time (T_R) to allow the current to stabilize; subsequently followed by an EIS test (Zhou, Birbilis et al. 2010). During the DC potential signal, the current at each potential is also recorded. Figure 10 demonstrates this process.

This method is used to assess the kinetic stability of electrochemical reactions over a range of potentials and pH, which gives a rather detailed insight into the corrosion behavior. In SPEIS however, by setting a range of voltage allows more details observation of how the system response at a particular voltage or 'step', even though it is not sufficient to explain the kinetic processes occurring. As reported by Zhou, SPEIS is able to illustrate the effect of pH and potential on the corrosion kinetics with clarity and able to obtain more information beyond the Pourbaix diagram (Zhou, Birbilis et al. 2010), including structural effects such as alloying, etc.

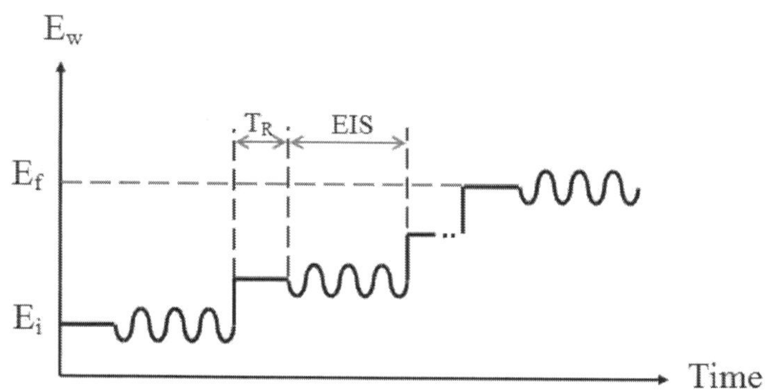

Figure 10. Illustration of polarizing signal during SPEIS

5.4. Potentiostatic Transients For Determination Of Metastable And Stable Pitting

Since pitting is the most common type of corrosion in Al alloys, researchers commonly default to the investigation of pitting potential, E_{pit} as a means for evaluating the corrosion performance. A more noble E_{pit} is often accepted as better resistance to pitting. However, there are some significant limitations in such simple assessments (Gupta, Sukiman et al. 2012). For example, the pitting potential yields no information regarding how many pits form or how large pits may be. Additionally, the environment plays a key role in the severity of pitting damage (i.e. pit depth, pit size) (Cavanaugh 2009).

Pits that form at pitting potential are known as stable pits. However, at potentials slightly below pits stabilization (at E_{pit}), pitting like events occur that are known as metastable pit (small surges of current that repassivate). Wu suggested that individual metastable pitting events influence subsequent events (Wu, Scully et al. 1997), and such events are readily observed in Al-alloys. Figure 11 indicates the metastable pitting region in a typical potentiodynamic polarisation curve, where E_{pit} is also depicted.

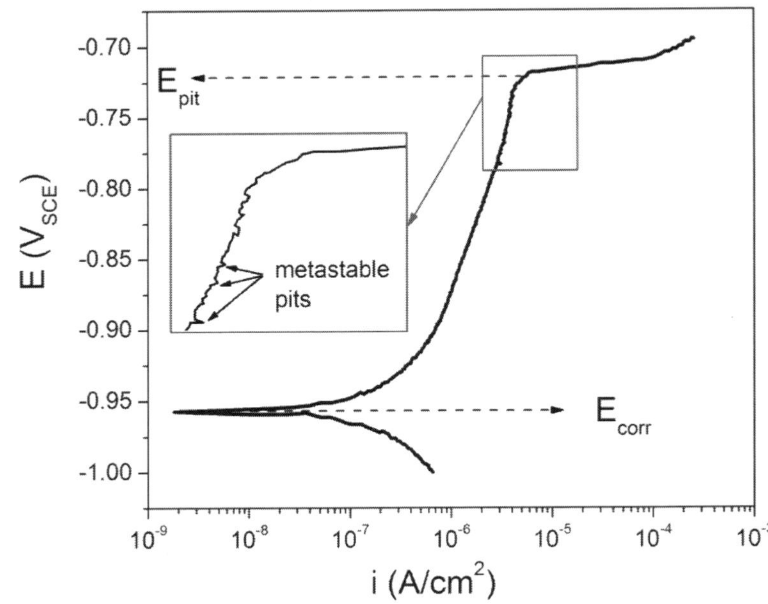

Figure 11. Polarisation curve indicating the corrosion potential E_{corr} and E_{pit} with metastable pits region prior to E_{pit} highlighted in the inset (adapted from (Cavanaugh 2009))

In order to measure metastable pitting, a potentiostatic value is selected at a potential just below $E_{pit}$where the metastable pitting events are the most frequent (while still in the passive region) (Cavanaugh 2009). Metastable pitting events are measured by counting the current fluctuations when an alloy is held potentiostatically for a period of time. The transient currents as shown in Figure 12 depend on the nucleation, growth and repassivation of the metastable pits.

The number of metastable pitting is expected to be proportional to the amount of pits formed (Williams, Stewart et al. 1994; Cavanaugh, Birbilis et al. 2012). This is confirmed by (Gupta, Sukiman et al. 2012), where the numbers of metastable pitting from potentiostatic test for various commercial Al alloys correlate with stable pitting tested in a long term immersion test. Ilevbare and Burstein however, stated that multiple metastable events may correspond to only one pit site (Burstein, Liu et al. 2004; Ilevbare, Schneider et al. 2004; Sasaki and Isaacs 2004; Trueman 2005; Speckert and Burstein 2011), however most of such analyses were not performed on Al-alloys. Since there are few studies on metastable pitting of Al alloy, the associated theory and principles will evolve in the coming decade. It is however obvious that metastable pitting analysis can be used to compare the pitting susceptibility between different environments

[Cavanaugh 2009] and alloy systems [Gupta, Sukiman et al. 2012] more effectively than an examination of (E_{pit}) alone.

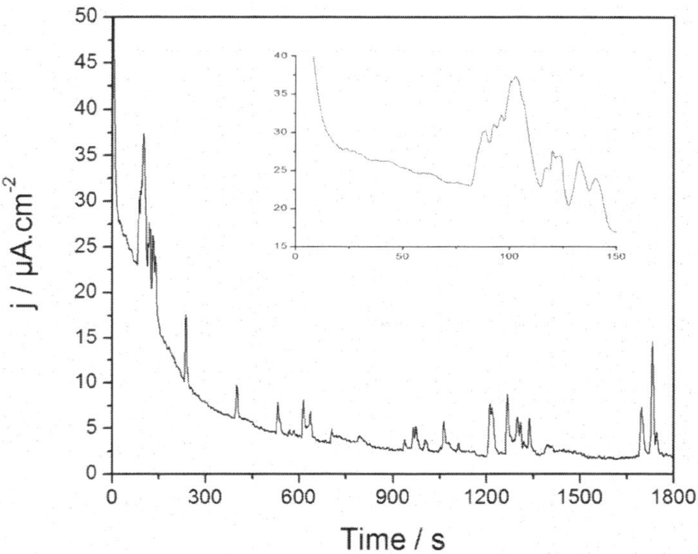

Figure 12. Current transient for AA5083-H116 obtained at 25mV below E_{pit}. Inset shows a zoom of a region of interest that represents the typical transient features (adapted from(Gupta, Sukiman et al. 2012))

Aside from the methods described above, the use of modern tests does not exclude the use of other traditional tests that require longer testing times and samples such as salt fog spray tests, weight determination, outdoor exposure, or cyclic salt fog-climatic chamber tests, but they aim at providing unique insights that are relevant to specific problems at hand such as pitting germination and inhibitor selection.

It is also worth mentioning here, the great interest that local electrochemical techniques have recently attracted to evaluate the corrosion protection offered when the coatings are damaged (i.e. self-healing). These techniques have been successfully applied to several conceptually different self-healing systems like scanning vibrating electrode technique (SVET) for encapsulated agents (Hughes, Cole et al. 2010;García, Fischer et al. 2011), scanning electrochemical microscope (SECM) for shape memory polymers, selective ion electrode technique (SIET) to evaluate activity of metals (Lamaka, Karavai et al. 2008), and local electrochemical impedance spectroscopy (LEIS) for evaluation of corrosion protection by inhibitor release from coatings (Jorcin, Aragon et al. 2006)

6. SUMMARY AND CHALLENGES

This chapter has given an abridged and focused treatment of the corrosion behavior of Al alloys on the basis of chemistry, microstructure and environment. Understanding these parameters is crucial in deployment of Al-alloys, but also in the development of more durable Al-alloys. It is obvious that alloying (i.e. chemistry) and microstructure dictates not only mechanical strength – but also corrosion performance. In this chapter we stopped short of discussing corrosion propagation such as intergranular corrosion, exfoliation corrosion and stress corrosion cracking – given space constraints, however such topics (along with hydrogen embrittlement) remain critical for structural alloy deployment.

In regards to corrosion protection, we have attempted to cover some modern developments and present some new techniques to assess the corrosion behavior designed to meet the more complicated and challenging requirements of inhibitor selection / chromate replacement. The information gathered from these techniques is beneficial for future protection and alloy development.

Looking to the future, the current strategy to increase strength by precipitation hardening of a crystalline matrix may no longer be feasible if an alloy with strength higher than 1000MPa is required. Methods to meet this threshold include (the un-upscalable) severe plastic deformation method such as High Pressure Torsion (Liddicoat, Liao et al. 2010) – however the thermal stability of these structures is low. Alternatively the strength of Al based metals could be significantly enhanced (up to 1500MPa) when rapidly quenched to form amorphous alloys (Masumoto 1994; Inoue and Takeuchi 2004; Li, Li et al. 2009; Yang, Yao et al. 2009). Such alloys typically use transition metals such as Zr, Ti, Nb, La, etc (Inoue, Gook et al. 1995; Zawrah and Shaw 2003; Rizzi and Battezzati 2004; Samanta, Manna et al. 2007; Huang, Li et al. 2008; Li, Li et al. 2009), however, again – the stability of such structures is unknown, and the ability form large components is also a challenge. From a corrosion point of view, amorphous Al-alloys appear promising (Manna, Chattopadhyay et al. 2004; Lucente and Scully 2007; Lucente and Scully 2008; Tailleart, Huang et al. 2012) with the corrosion behavior of such alloys is not widely explored therefore leaves a lot of opportunities for new discoveries.

REFERENCES

1. Aballe, A., M. Bethencourt, et al. (2001). "Localized alkaline corrosion of alloy AA5083 in neutral 3.5% NaCl solution." Corrosion Science 43(9): 1657-1674.

2. Allahar, K. N., V. Upadhyay, et al. (2010). "Characterizing the relaxation of the open-circuit potential during an AC-DC-AC accelerated test." Corrosion 66(9): 0950011-09500111.

3. Allahar, K. N., D. Wang, et al. (2009). Real time monitoring of an air force topcoat/mg-rich primer system in b117 exposure by -embedded electrodes.

4. Ambat, R., A. J. Davenport, et al. (2006). "Effect of iron-containing intermetallic particles on the corrosion behaviour of aluminium." Corrosion Science 48(11): 3455-3471.

5. Ambat, R. and E. S. Dwarakadasa (1992). "The Influence of PH on the Corrosion of Medium Strength Aerospace Alloy-8090, Alloy-2091 And Alloy-2014." Corrosion Science 33(5): 681-690.

6. Anawati, S. Diplas, et al. (2011). "Effect of copper on anodic activity of aluminum-lead model alloy in chloride solution." Journal of The Electrochemical Society 158(5): C158-C163.

7. Anawati, S. Diplas, et al. (2011). "Surface Characterization of Heat Treated AlPbCu Model Alloys." Journal of The Electrochemical Society 158(6): C178-C184.

8. Anawati, B. Graver, et al. (2010). "Multilayer corrosion of aluminum activated by lead." Journal of The Electrochemical Society 157(10): C313-C320.

9. Andreatta, F., H. Terryn, et al. (2004). "Corrosion behaviour of different tempers of AA7075 aluminium alloy." Electrochimica Acta 49(17–18): 2851-2862.

10. Ashtari, P., H. Tezuka, et al. (2003). "Influence of Sr and Mn additions on intermetallic compound morphologies in Al-Si-Cu-Fe cast alloys." Materials Transactions 44(12): 2611-2616.

11. Baer, D. R., C. F. Windisch, et al. (2000). "Influence of Mg on the corrosion of Al." Journal of Vacuum Science & Technology a-Vacuum Surfaces and Films 18(1): 131-136.

12. Barsoukov, E. and J. R. Macdonald (2005). Solid State Devices, in Impedance spectroscopy: theory, experiment, and applications, Wiley-Interscience, Inc.

13. Battocchi, D., A. M. Simoes, et al. (2006). "Comparison of testing solutions on the protection of Al-alloys using a Mg-rich primer." Corrosion Science 48(8): 2226-2240.

14. Bethencourt, M., F. J. Botana, et al. (2004). "Lifetime prediction of waterborne acrylic paints with the AC-DC-AC method." Progress in Organic Coatings 49(3): 275-281.

15. Birbilis, N. and R. G. Buchheit (2005). "Electrochemical Characteristics of Intermetallic Phases in Aluminum Alloys." Journal of The Electrochemical Society 152(4): B140.

16. Birbilis, N., R. G. Buchheit, et al. (2005). "Inhibition of AA2024-T3 on a phase-by-phase basis using an environmentally benign inhibitor, cerium dibutyl phosphate." Electrochemical and Solid State Letters 8(11): C180-C183.

17. Birbilis, N., M. K. Cavanaugh, et al. (2006). "Electrochemical behavior and localized corrosion associated with Al7Cu2Fe particles in aluminum alloy 7075-T651." Corrosion Science 48(12): 4202-4215.

18. Birbilis, N., T. Muster, et al. (2011). Corrosion of Aluminum Alloys. Corrosion Mechanisms in Theory and Practice, Third Edition, CRC Press: 705-736.

19. Blanc, C., B. Lavelle, et al. (1997). "The role of precipitates enriched with copper on the susceptibility to pitting corrosion of the 2024 aluminium alloy." Corrosion Science 39(3): 495-510.

20. Blin, F., P. Koutsoukos, et al. (2007). "The corrosion inhibition mechanism of new rare earth cinnamate compounds - Electrochemical studies." Electrochimica Acta 52(21): 6212-6220.

21. Boag, A., A. E. Hughes, et al. (2011). "Corrosion of AA2024-T3 Part I: Localised corrosion of isolated IM particles." Corrosion Science 53(1): 17-26.

22. Boag, A., A. E. Hughes, et al. (2009). "How complex is the microstructure of AA2024-T3?" Corrosion Science 51(8): 1565-1568.

23. Bohm, S., H. N. McMurray, et al. (2001). "Novel environment friendly corrosion inhibitor pigments based on naturally occurring clay minerals." Materials and Corrosion-Werkstoffe Und Korrosion 52(12): 896-903.

24. Bon, S. A. F., S. D. Mookhoek, et al. (2007). "Route to stable non-spherical emulsion droplets." European Polymer Journal 43(11): 4839-4842.

25. Brunner, J. G., J. May, et al. (2010). "Localized corrosion of ultrafine-grained Al–Mg model alloys." Electrochimica Acta 55(6): 1966-1970.

26. Buchheit, R. G. (1995). "A Compilation of Corrosion Potentials Reported for Intermetallic Phases in Aluminum-Alloys." Journal of The Electrochemical Society 142(11): 3994-3996.

27. Buchheit, R. G. (2000). "Electrochemistry of θ (Al2Cu), S (Al2CuMg) and T1 (Al2CuLi) and localized corrosion and environment assisted cracking in high strength Al alloys." Materials Science Forum 331: II/.

28. Buchheit, R. G. and N. Birbilis (2010). "Electrochemical microscopy: An approach for understanding localized corrosion in microstructurally complex metallic alloys." Electrochimica Acta 55(27): 7853-7859.

29. Buchheit, R. G., R. P. Grant, et al. (1997). "Local Dissolution Phenomena Associated with S Phase (Al2CuMg) Particles in Aluminum Alloy 2024-T3." Journal of The Electrochemical Society 144(8): 2621-2628.

30. Buchheit, R. G., H. Guan, et al. (2003). "Active corrosion protection and corrosion sensing in chromate-free organic coatings." Progress in Organic Coatings 47(3-4): 174-182.

31. Buchheit, R. G., Hughes, A.E. (2003). Chromate and Chromate-Free Coatings. Corrosion: Fundamentals, Testing and Protection. C. Moosbrugger. Mterials Park, Oh, USA, ASM International. 13A: 720 - 735.

32. Buchheit, R. G., M. A. Martinez, et al. (2000). "Evidence for Cu Ion Formation by Dissolution and Dealloying the Al2CuMg Intermetallic Compound in Rotating Ring-Disk Collection Experiments." Journal of The Electrochemical Society 147(1): 119-124.

33. Buchheit, R. G., L. P. Montes, et al. (1999). "The Electrochemical Characteristics of Bulk-Synthesized Al[sub 2]CuMg." Journal of The Electrochemical Society 146(12): 4424-4428.

34. Buchheit, R. G., F. D. Wall, et al. (1995). "Anodic Dissolution-Based Mechanism for the Rapid Cracking, Preexposure Phenomenon Demonstrated by Aluminum-Lithium-Copper Alloys." Corrosion 51(6): 417-428.

35. Büchler, M., T. Watari, et al. (2000). "Investigation of the initiation of localized corrosion on aluminum alloys by using fluorescence microscopy." Corrosion Science 42(9): 1661-1668.

36. Buis, A. and J. Schijve (1992). "Stress-Corrosion Cracking Behavior of Al Li-2090-T83 In Artificial Seawater." Corrosion 48(11): 898-909.

37. Burstein, G. T., C. Liu, et al. (2004). "Origins of pitting corrosion." Corrosion Engineering Science and Technology 39(1): 25-30.

38. Campestrini, P., E. P. M. van Westing, et al. (2000). "Relation between microstructural aspects of AA2024 and its corrosion behaviour investigated using AFM scanning potential technique." Corrosion Science 42(11): 1853-1861.

39. Carroll, M. C., P. I. Gouma, et al. (2001). "Effects of minor Cu additions on a Zn-modified Al-5083 alloy." Materials Science and Engineering: A 319–321(0): 425-428.

40. Carroll, M. C., P. I. Gouma, et al. (2000). "Effects of Zn additions on the grain boundary precipitation and corrosion of Al-5083." Scripta Materialia 42(4): 335-340.

41. Cavanaugh, M., N. Birbilis, et al. (2007). "Investigating localized corrosion susceptibility arising from Sc containing intermetallic Al3Sc in high strength Al-alloys." Scripta Materialia 56(11): 995-998.

42. Cavanaugh, M. K. (2009). Modelling the environmental dependence of localized corrosion evolution in AA7075-T651, Ohio State University.

43. Cavanaugh, M., N. Birbilis, et al. (2007). "Investigating localized corrosion susceptibility arising from Sc containing intermetallic Al3Sc in high strength Al-alloys." Scripta Materialia 56(11): 995-998.

44. Cavanaugh, M. K., N. Birbilis, et al. (2009). "A Quantitative Study on the Effects of Environment and Microstructure on Pit Initiation in Al-alloys." ECS Transactions 16(52): 1-11.

45. Cavanaugh, M. K., N. Birbilis, et al. (2012). "Modeling pit initiation rate as a function of environment for Aluminum alloy 7075-T651." Electrochimica Acta 59: 336-345.

46. Cavanaugh, M. K., R. G. Buchheit, et al. (2009). "Evaluation of a simple microstructural-electrochemical model for corrosion damage accumulation in microstructurally complex aluminum alloys." Engineering Fracture Mechanics 76(5): 641-650.

47. Cavanaugh, M. K., R. G. Buchheit, et al. (2010). "Modeling the environmental dependence of pit growth using neural network approaches." Corrosion Science 52(9): 3070-3077.

48. Chambers, B. D. and S. R. Taylor (2007). "High-Throughput Assessment of Inhibitor Synergies on Aluminum Alloy 2024-T3 Through Measurement of Surface Copper Enrichment." Corrosion 63(3): 268-276.

49. Chambers, B. D., S. R. Taylor, et al. (2005). "Rapid Discovery of Corrosion Inhibitors and Synergistic Combinations Using High-Throughput Screening Methods." Corrosion 61(5): 480-489.

50. Chen, G. S., M. Gao, et al. (1996). "Microconstituent-Induced Pitting Corrosion in Aluminum Alloy 2024-T3." Corrosion (Houston) 52(1): 8-15.

51. Cho, S. H., S. R. White, et al. (2009). "Self-Healing Polymer Coatings." Advanced Materials 21(6): 645-+.

52. Davenport, A. J., Y. Yuan, et al. (2006). "Intergranular Corrosion and Stress Corrosion Cracking of Sensitised AA5182." Materials Science Forum 519-521: 641-646.

53. Davis, J. R. (1999). Corrosion of aluminum and aluminum alloys, Materials Park, OH : ASM International.

54. Deacon, G. B., M. Forsyth, et al. (2009). "Synthesis and Characterisation of Rare Earth Complexes Supported by para-Substituted Cinnamate Ligands." Zeitschrift Fur Anorganische Und Allgemeine Chemie 635(6-7): 833-839.

55. Dorward, R. C. and T. R. Pritchett (1988). "Advanced aluminium alloys for aircraft and aerospace applications." Materials & Design 9(2): 63-69.

56. Dry, C. (1996). "Procedures developed for self-repair of polymer matrix composite materials." Composite Structures 35(3): 263-269.

57. Du, Y. J., M. Damron, et al. (2001). "Inorganic/organic hybrid coatings for aircraft aluminum alloy substrates." Progress in Organic Coatings 41(4): 226-232.

58. Eckermann, F., T. Suter, et al. (2008). "The influence of MgSi particle reactivity and dissolution processes on corrosion in Al–Mg–Si alloys." Electrochimica Acta 54(2): 844-855.

59. Eidhed, W. (2008). "Effects of solution treatment time and Sr-modification on microstructure and mechanical property of Al-Si piston alloy." Journal of Materials Science and Technology 24(1): 29-32.

60. Fang, H. C., K. H. Chen, et al. (2009). "Effect of Cr, Yb and Zr additions on localized corrosion of Al–Zn–Mg–Cu alloy." Corrosion Science 51(12): 2872-2877.

61. Ferrer, C. P., M. G. Koul, et al. (2003). "Improvements in strength and stress corrosion cracking properties in aluminum alloy 7075 via low-temperature retrogression and re-aging heat treatments." Corrosion 59(6): 520-528.

62. Fischer, H. R. (2010). natural Science 2: 873-901.

63. Fleck, P., D. Calleros, et al. (2000). "Retrogression and reaging of 7075 T6 aluminum alloy." Materials Science Forum 331: I/.

64. Forsyth, M., T. Markley, et al. (2008). "Inhibition of corrosion on AA2024-T3 by new environmentally friendly rare earth organophosphate compounds." Corrosion 64(3): 191-197.

65. Frankel, G. S. (1998). "Pitting Corrosion of Metals." Journal of The Electrochemical Society 145(6): 2186-2198.

66. Frankel, G. S., A. J. Davenport, et al. (1992). "X-ray absorption study of electrochemically grown oxide films on Al-Cr sputtered alloys." Journal of The Electrochemical Society 139(7): 1812-1820.

67. Frankel, G. S., M. A. Russak, et al. (1989). "Pitting of Sputtered Aluminum Alloy Thin Films." Journal of The Electrochemical Society 136(4): 1243-1244.

68. Fuller, C. B., A. R. Krause, et al. (2002). "Microstructure and mechanical properties of a 5754 aluminum alloy modified by Sc and Zr additions." Materials Science and Engineering A 338(1-2): 8-16.

69. Furman, S. A., F. H. Scholes, et al. (2006). "Corrosion in artificial defects. II. Chromate reactions." Corrosion Science 48(7): 1827-1847.

70. Galvele, J. R. (1976). "Transport Processes and the Mechanism of Pitting of Metals." Journal of The Electrochemical Society 123(4): 464-474.

71. García, S. J., H. R. Fischer, et al. (2011). "Self-healing anticorrosive organic coating based on an encapsulated water reactive silyl ester:

Synthesis and proof of concept." Progress in Organic Coatings 70(2-3): 142-149.

72. García, S. J., Mol, J.M.C., Muster, T.H., Hughes, A.E., Mardel, J., Miller, T., Markely, T., Terryn, H., de Wit, J.H.W. (2011). Advances in the Selection and use of Rare-Earth-Based Inhibitors for Self Healing Organic Coatings, Accepted for publication in Self-Healing Properties of New Surface Treatments. Green Inhibitors. L. Fedrizzi, EFC-Maney Publishing. 58.

73. García, S. J., T. H. Muster, et al. (2010). "The influence of pH on corrosion inhibitor selection for 2024-T3 aluminium alloy assessed by high-throughput multielectrode and potentiodynamic testing." Electrochimica Acta 55(7): 2457-2465.

74. García, S. J., M. T. Rodríguez, et al. (2007). "Evaluation of cure temperature effects in cataphoretic automotive primers by electrochemical techniques." Progress in Organic Coatings 60(4): 303-311.

75. García, S. J. and J. Suay (2006). "Anticorrosive properties of an epoxy-Meldrum acid cured system catalyzed by erbium III trifluromethanesulfonate." Progress in Organic Coatings 57(4): 319-331.

76. García, S. J. and J. Suay (2006). "Application of electrochemical techniques to study the effect on the anticorrosive properties of the addition of ytterbium and erbium triflates as catalysts on a powder epoxy network." Progress in Organic Coatings 57(3): 273-281.

77. García, S. J. and J. Suay (2007). "A comparative study between the results of different electrochemical techniques (EIS and AC/DC/AC). Application to the optimisation of the cataphoretic and curing parameters of a primer for the automotive industry." Progress in Organic Coatings 59(3): 251-258.

78. García, S. J. and J. Suay (2007). "Influence on the anticorrosive properties of the use of erbium (III) trifluoromethanesulfonate as initiator in an epoxy powder clearcoat." Corrosion Science 49(8): 3256-3275.

79. García, S. J. and J. Suay (2009). "Optimization of deposition voltage of cataphoretic automotive primers assessed by EIS and AC/DC/AC." Progress in Organic Coatings 66(3): 306-313.

80. Garrard, W. N. (1994). "Corrosion Behavior of Aluminum-Lithium Alloys." Corrosion 50(3): 215-225.

81. Gimenez, P., J. J. Rameau, et al. (1981). "Experimental pH potential diagram of aluminium for seawater." Corrosion 37: 673-682.

82. Giummarra, C., B. Thomas, et al. (2007). "New Al-Li Alloys for Aerospace Applications." Proceedings of the Light Metals Technology Conference.

83. Goswami, R., G. Spanos, et al. (2010). "Precipitation behavior of the β phase in Al-5083." Materials Science and Engineering: A 527(4-5): 1089-1095.

84. Graver, B., A. M. Pedersen, et al. (2009). "Anodic Activation of Aluminum by Trace Element Tin." ECS Transactions 16(52): 55-69.

85. Graver, B., A. T. J. van Helvoort, et al. (2010). "Effect of heat treatment on anodic activation of aluminium by trace element indium." Corrosion Science 52(11): 3774-3781.

86. Guillaumin, V. and G. Mankowski (1998). "Localized corrosion of 2024 T351 aluminium alloy in chloride media." Corrosion Science 41(3): 421-438.

87. Guillaumin, V. and G. Mankowski (2000). "Localized corrosion of 6056 T6 aluminium alloy in chloride media." Corrosion Science 42(1): 105-125.

88. Gundersen, J. T. B., A. Aytaç, et al. (2004). "Effect of heat treatment on electrochemical behaviour of binary aluminium model alloys." Corrosion Science 46(3): 697-714.

89. Gupta, R. K., N. L. Sukiman, et al. (2012). "Metastable pitting characteristics of aluminium alloys measured using current transients during potentiostatic polarisation." Electrochimica Acta 66: 245-254.

90. Gupta, R. K., N. L. Sukiman, et al. (2012). "Electrochemical Behavior and Localized Corrosion Associated with Mg2Si Particles in Al and Mg Alloys." ECS Electrochemistry Letters 1(1): B1-B3.

91. Harvey, T. G., S. G. Hardin, et al. (2011). "The effect of inhibitor structure on the corrosion of AA2024 and AA7075." Corrosion Science 53(6): 2184-2190.

92. Harvey, T. G., A. E. Hughes, et al. (2008). "Non-chromate deoxidation of AA2024-T3: Sodium bromate–nitric acid (20–60)." Applied Surface Science 254(11): 3562-3575.

93. Hatch, J. E. (1984). Aluminum: properties and physical metallurgy, Metals Park, Ohio : American Society for Metals.

94. He, Y., X. Zhang, et al. (2010). "Effect of minor Cr, Mn, Zr, Ti and B on grain refinement of as-cast Al-Zn-Mg-Cu alloys." Xiyou Jinshu Cailiao Yu Gongcheng/Rare Metal Materials and Engineering 39(7): 1135-1140.

95. Hinton, B. R. W., N. Dubrule, et al. (2006). Raman, EDS and SEM studies of the interaction of corrosion inhibitor Ce(dbp)3 with AA2024-T3. 4th International Symposium on Aluminium Surface Science and Technology. Beaune, France.

96. Hirth, S. M., G. J. Marshall, et al. (2001). "Effects of Si on the aging behaviour and formability of aluminium alloys based on AA6016." Materials Science and Engineering A 319-321: 452-456.

97. Ho, D., N. Brack, et al. (2006). "Cerium dibutylphosphate as a corrosion inhibitor for AA2024-T3 aluminum alloys." Journal of the Electrochemical Society 153(9): B392-B401.

98. Hollaender, J. (1997). "Rapid assessment of food/package interactions by electrochemical impedance spectroscopy (EIS)." Food Additives and Contaminants 14(6-7): 617-626.

99. Huang, Z. H., J. F. Li, et al. (2008). "Primary crystallization of Al–Ni–RE amorphous alloys with different type and content of RE." Materials Science and Engineering: A 489(1–2): 380-388.

100. Hughes, A. E., A. Boag, et al. (2011). "Corrosion of AA2024-T3 Part II: Co-operative corrosion." Corrosion Science 53(1): 27-39.

101. Hughes, A. E., A. M. Glenn, et al. (2012). A Consistent Description of Intermetallic Particle Composition: An analysis of 10 Batches of AA2024-T3. Aluminium Surface Science & Technology, Sorrento, Italy.

102. Hughes, A. E., C. MacRae, et al. (2010). "Sheet AA2024-T3: A new investigation of microstructure and composition." Surface and Interface Analysis 42(4): 334-338.

103. Hughes, A. E., I. S. Cole, et al. (2010). "Designing green, self-healing coatings for metal protection." NPG Asia Materials 2(4): 143-151.

104. Hughes, A. E., I. S. Coles, et al. (2010). "Combining Green and Self Healing for a new Generation of Coatings for Metal Protection." Nature Asia Materials 2(4): 143-151.

105. Ilevbare, G. O., O. Schneider, et al. (2004). "In Situ Confocal Laser Scanning Microscopy of AA 2024-T3 Corrosion Metrology." Journal of The Electrochemical Society 151(8): B453.

106. Inoue, A., J. S. Gook, et al. (1995). "New amorphous alloys in Al-Mg-Ln (Ln = La, Ce or Nd) systems prepared by rapid solidification (rapid publication)." Materials Transactions, JIM 36(7): 794-796.

107. Inoue, A. and A. Takeuchi (2004). "Recent progress in bulk glassy, nanoquasicrystalline and nanocrystalline alloys." Materials Science and Engineering: A 375–377(0): 16-30.

108. Jain, S. (2006). Corrosion and protection of heterogeneous cast Al-Si (356) and Al-Si-Fe-Cu (380) alloys by chromate and cerium inhibitors, Ohio State University.

109. Jain, S., M. L. C. Lim, et al. (2012). "Spreading of intergranular corrosion on the surface of sensitized Al-4.4Mg alloys: A general finding." Corrosion Science 59: 136-147.

110. Jia, Z., B. Graver, et al. (2008). "Effect of magnesium on segregation of trace element lead and anodic activation in aluminum Alloys." Journal of The Electrochemical Society 155(1): C1-C7.

111. Jones, R. H., D. R. Baer, et al. (2001). "Role of Mg in the stress corrosion cracking of an Al-Mg alloy." Metallurgical and Materials Transactions A: Physical Metallurgy and Materials Science 32(7): 1699-1711.

112. Jorcin, J. B., E. Aragon, et al. (2006). "Delaminated areas beneath organic coating: A local electrochemical impedance approach." Corrosion Science 48(7): 1779-1790.

113. Joshua Du, Y., M. Damron, et al. (2001). "Inorganic/organic hybrid coatings for aircraft aluminum alloy substrates." Progress in Organic Coatings 41(4): 226-232.

114. Juffs, L. (2002). Investigation of Corrosion Coating Deposition on Microscopic and Macroscopic Intermetallic Phases of Aluminium Alloys. Master of Science, RMIT.

115. Juffs, L., A. E. Hughes, et al. (2002). "The use of macroscopic modelling of intermetallic phases in aluminium alloys in the study of ferricyanide accelerated chromate conversion coatings." Corrosion Science 44(8): 1755-1781.

116. Juffs, L., A. E. Hughes, et al. (2001). "The use of macroscopic modelling of intermetallic phases in aluminium alloys in the study of ferricyanide accelerated chromate conversion coatings." Micron 32(8): 777-787.

117. Kallip, S., A. C. Bastos, et al. (2010). "A multi-electrode cell for high-throughput SVET screening of corrosion inhibitors." Corrosion Science 52(9): 3146-3149.

118. Kannan, M. B. and V. S. Raja (2010). Enhancing the localized corrosion resistance of high strength 7010 Al-alloy. 138: 1-6.

119. Kendig, M. W. and R. G. Buchheit (2003). "Corrosion Inhibition of Aluminum and Aluminum Alloys by Soluble Chromates, Chromate Coatings, and Chromate-Free Coatings." Corrosion 59(5): 379-400.

120. Keuong, Y. W., J. H. Nordlien, et al. (2003). "Electrochemical activation of aluminum by trace element lead." Journal of The Electrochemical Society 150(11): B547-B551.

121. Khramov, A. N., N. N. Voevodin, et al. (2004). "Hybrid organo-ceramic corrosion protection coatings with encapsulated organic corrosion inhibitors." Thin Solid Films 447–448(0): 549-557.

122. Kim, K. T., J. M. Kim, et al. (2005). Effect of alloying elements on the strength and casting characteristics of high strength Al-Zn-Mg-Cu alloys. 475-479: 2539-2542.

123. Kim, W. J., C. S. Chung, et al. (2003). "Optimization of strength and ductility of 2024 Al by equal channel angular pressing (ECAP) and post-ECAP aging." Scripta Materialia 49(4): 333-338.

124. Koroleva, E. v., G. e. Thompson, et al. (1999). "Surface morphological changes of aluminium alloys in alkaline solution:: effect of second phase material." Corrosion Science 41(8): 1475-1495.

125. Lamaka, S. V., O. V. Karavai, et al. (2008). "Monitoring local spatial distribution of Mg2+, pH and ionic currents." Electrochemistry Communications 10(2): 259-262.

126. Larsen, M. H., J. C. Walmsley, et al. (2008). "Intergranular corrosion of copper-containing AA6xxx AlMgSi aluminum alloys." Journal of The Electrochemical Society 155(11): C550-C556.

127. Lavernia, E. J. and N. J. Grant (1987). "Aluminum Lithium Alloys." Journal of Materials Science 22(5): 1521-1529.

128. Leblanc, P. and G. S. Frankel (2002). "A study of corrosion and pitting initiation of AA2024-T3 using atomic force microscopy." Journal of The Electrochemical Society 149(6): B239-B247.

129. Lee, Y. B., D. H. Shin, et al. (2004). "Effect of annealing temperature on microstructures and mechanical properties of a 5083 Al alloy deformed at cryogenic temperature." Scripta Materialia 51(4): 355-359.

130. Li, C., D. Y. Li, et al. (2009). "Microstructure and mechanical properties of multicomponent aluminum alloy by rapid solidification." Journal of Materials Engineering and Performance 18(1): 79-82.

131. Li, J. F., Z. Q. Zheng, et al. (2007). "Exfoliation corrosion and electrochemical impedance spectroscopy of an Al-Li alloy in EXCO solution." Materials and Corrosion 58(4): 273-279.

132. Liddicoat, P. V., X.-z. Liao, et al. (2010). "Nanostructural hierarchy increases the strength of aluminium alloys." Nature Communications 1(6): 63-63.

133. Lin, J. C., H. L. Liao, et al. (2006). "Effect of heat treatments on the tensile strength and SCC-resistance of AA7050 in an alkaline saline solution." Corrosion Science 48(10): 3139-3156.

134. Liu, Y. and Y. F. Cheng (2010). "Role of second phase particles in pitting corrosion of 3003 Al alloy in NaCl solution." Materials and Corrosion 61(3): 211-217.

135. Liu, Y. and Y. F. Cheng (2011). "Characterization of passivity and pitting corrosion of 3003 aluminum alloy in ethylene glycol-water solutions." Journal of Applied Electrochemistry 41(2): 151-159.

136. Lucente, A. M. and J. R. Scully (2007). "Pitting of Al-based amorphous-nanocrystalline alloys with solute-lean nanocrystals." Electrochemical and Solid-State Letters 10(5): 39-43.

137. Lucente, A. M. and J. R. Scully (2008). "Localized corrosion of Al-based amorphous-nanocrystalline alloys with solute-lean nanocrystals:

Pit stabilization." Journal of The Electrochemical Society 155(5): C234-C243.

138. Lynch, S. P., S. P. Knight, et al. (2009). Stress-corrosion cracking of Al-Zn-Mg-Cu alloys effects of composition and heat-treatment.

139. Mahajanarn, S. P. V. and R. G. Buchheit (2008). "Characterization of inhibitor release from Zn-Al- V10O28 (6-) hydrotalcite pigments and corrosion protection from hydrotalcite-pigmented epoxy coatings." Corrosion 64(3): 230-240.

140. Manna, I., P. P. Chattopadhyay, et al. (2004). "Development of amorphous and nanocrystalline Al65Cu35−xZrx alloys by mechanical alloying." Materials Science and Engineering: A 379(1–2): 360-365.

141. Mardel, J., S. J. Garcia, et al. (2011). "The characterisation and performance of Ce(dbp)3-inhibited epoxy coatings." Progress in Organic Coatings 70(2–3): 91-101.

142. Markley, T. A., M. Forsyth, et al. (2007). "Corrosion protection of AA2024-T3 using rare earth diphenyl phosphates." Electrochimica Acta 52(12): 4024-4031.

143. Markley, T. A., A. E. Hughes, et al. (2007). "Influence of praseodymium - Synergistic corrosion inhibition in mixed rare-earth diphenyl phosphate systems." Electrochemical and Solid State Letters 10(12): C72-C75.

144. Marlaud, T., A. Deschamps, et al. (2010). "Evolution of precipitate microstructures during the retrogression and re-ageing heat treatment of an Al-Zn-Mg-Cu alloy." Acta Materialia 58(14): 4814-4826.

145. Martin, J. W. (1988). "Aluminum-Lithium Alloys." Annual Review of Materials Science 18: 101-119.

146. Masumoto, T. (1994). "Recent progress in amorphous metallic materials in Japan." Materials Science and Engineering: A 179–180, Part 1(0): 8-16.

147. Mazurkiewicz, B. and A. Piotrowski (1983). "The electrochemical behaviour of the Al2Cu intermetallic compound." Corrosion Science 23(7): 697-707.

148. McCafferty, E. (2010). Passivity

149. Introduction to Corrosion Science, Springer New York: 209-262.

150. McKenzie, P. W. J. and R. Lapovok (2010). "ECAP with back pressure for optimum strength and ductility in aluminium alloy 6016. Part 2: Mechanical properties and texture." Acta Materialia 58(9): 3212-3222.

151. Medco.

152. Mondolfo, L. F. (1971). "Discussion of "grain-size refining of primary crystals in hypereutectic Al-Si and Al-Ge alloys"." Metallurgical Transactions 2(4): 1254.

153. Mondolfo, L. F. and J. G. Barlock (1975). "Effect of superheating on structure of some aluminum alloys." Metallurgical Transactions B 6(4): 565-572.

154. Mookhoek, S. D., H. R. Fischer, et al. (2009). "A numerical study into the effects of elongated capsules on the healing efficiency of liquid-based systems." Computational Materials Science 47(2): 506-511.

155. Mookhoek, S. D., S. C. Mayo, et al. (2010). "Applying SEM-Based X-ray Microtomography to Observe Self-Healing in Solvent Encapsulated Thermoplastic Materials." Advanced Engineering Materials 12(3): 228-234.

156. Muller, I. L. and J. R. Galvele (1977). "Pitting potential of high purity binary aluminium alloys—II. AlMg and AlZn alloys." Corrosion Science 17(12): 995-1007.

157. Muster, T. H., A. E. Hughes, et al. (2009). "A rapid screening multi-electrode method for the evaluation of corrosion inhibitors." Electrochimica Acta 54(12): 3402-3411.

158. Muster, T. H., A. E. Hughes, et al. (2009). Cu Distributions in Aluminium Alloys, New York, Nova Science Publishers.

159. Muster, T. H., Hughes, A.E., Thompson. G.E. (2009). Cu Distributions in Aluminium Alloys. New York, Nova Science Publishers.

160. Nisancioglu, K. (1990). "Electrochemical Behavior of Aluminum-Base Intermetallics Containing Iron." Journal of The Electrochemical Society 137(1): 69-77.

161. Nisancioglu, K. and H. Holtan (1978). "Measurement of the critical pitting potential of aluminium." Corrosion Science 18(9): 835-849.

162. Norova, M. T., I. N. Ganiev, et al. (2003). "Enhancement of the corrosion resistance of aluminum-lithium alloys by microalloying with calcium." Russian Journal of Applied Chemistry 76(4): 547-549.

163. Oguocha, I., O. Adigun, et al. (2008). "Effect of sensitization heat treatment on properties of Al–Mg alloy AA5083-H116." Journal of Materials Science 43(12): 4208-4214.

164. Oliveira Jr, A. F., M. C. de Barros, et al. (2004). "The effect of RRA on the strength and SCC resistance on AA7050 and AA7150 aluminium alloys." Materials Science and Engineering A 379(1-2): 321-326.

165. Orazem, M. E. and B. Tribollet (2008). Semiconducting Systems, in Electrochemical Impedance Spectroscopy, John Wiley & Sons.

166. Osborne, J. H., K. Y. Blohowiak, et al. (2001). "Testing and evaluation of nonchromated coating systems for aerospace applications." Progress in Organic Coatings 41(4): 217-225.

167. Park, J. O., C. H. Paik, et al. (1999). "Influence of Fe-Rich Intermetallic Inclusions on Pit Initiation on Aluminum Alloys in Aerated NaCl." Journal of The Electrochemical Society 146(2): 517-523.

168. Pedersen, L. and L. Arnberg (2001). "The effect of solution heat treatment and quenching rates on mechanical properties and microstructures in AlSiMg foundry alloys." Metallurgical and Materials Transactions A: Physical Metallurgy and Materials Science 32(3): 525-532.

169. Perrault, G. G. (1979). "Role of Hydrides in the Equilibrium of Aluminum in Aqueous Solutions." J Electrochem Soc 126(2): 199-204.

170. Poelman, M., M. G. Olivier, et al. (2005). "Electrochemical study of different ageing tests for the evaluation of a cataphoretic epoxy primer on aluminium." Progress in Organic Coatings 54(1): 55-62.

171. Polmear, I. J. (2006). Light alloys : from traditional alloys to nanocrystals, Oxford ; Burlington, MA : Elsevier/Butterworth-Heinemann.

172. Poole, W. J., J. A. Seter, et al. (2000). "A model for predicting the effect of deformation after solution treatment on the subsequent artificial aging behavior of AA7030 and AA7108 alloys." Metallurgical and

Materials Transactions A: Physical Metallurgy and Materials Science 31(9): 2327-2338.

173. Pourbaix, M. (1974). Atlas of Electrochemical Equilibria in Aqueous Solutions. National Association of Corrosion Engineers, Houston, TX, USA.

174. Premendra, P., H. Terryn, et al. (2009). "A comparative electrochemical study of commercial and model aluminium alloy (AA5050)." Materials and Corrosion 60(6): 399-406.

175. Puga, H., S. Costa, et al. (2011). "Influence of ultrasonic melt treatment on microstructure and mechanical properties of AlSi9Cu3 alloy." Journal of Materials Processing Technology 211(11): 1729-1735.

176. Ralston, K. D., N. Birbilis, et al. (2010). "Role of nanostructure in pitting of Al-Cu-Mg alloys." Electrochimica Acta 55(27): 7834-7842.

177. Ralston, K. D., N. Birbilis, et al. (2010). "Revealing the relationship between grain size and corrosion rate of metals." Scripta Materialia 63(12): 1201-1204.

178. Raps, D., T. Hack, et al. (2009). "Electrochemical study of inhibitor-containing organic-inorganic hybrid coatings on AA2024." Corrosion Science 51(5): 1012-1021.

179. Raviprasad, K., C. R. Hutchinson, et al. (2003). "Precipitation processes in an Al-2.5Cu-1.5Mg (wt. %) alloy microalloyed with Ag and Si." Acta Materialia 51(17): 5037-5050.

180. Ringer, S. P., K. Hono, et al. (1996). "Nucleation of precipitates in aged AlCuMg(Ag) alloys with high Cu:Mg ratios." Acta Materialia 44(5): 1883-1898.

181. Rizzi, P. and L. Battezzati (2004). "Mechanical properties of Al based amorphous and devitrified alloys containing different rare earth elements." Journal of Non-Crystalline Solids 344(1–2): 94-100.

182. Rodríguez, M. T., J. J. Gracenea, et al. (2004). "Testing the influence of the plasticizers addition on the anticorrosive properties of an epoxy primer by means of electrochemical techniques." Progress in Organic Coatings 50(2): 123-131.

183. Rosalbino, F., E. Angelini, et al. (2003). "Influence of the rare earth content on the electrochemical behaviour of Al-Mg-Er alloys." Intermetallics 11(5): 435-441.

184. Sævik, Ø., Y. Yu, et al. (2005). "Characterization of lead enrichment on electrochemically active AlPb model alloy." Journal of The Electrochemical Society 152(9): B334-B341.

185. Samadzadeh, M., S. H. Boura, et al. (2010). "A review on self-healing coatings based on micro/nanocapsules." Progress in Organic Coatings 68(3): 159-164.

186. Samanta, A., I. Manna, et al. (2007). "Phase evolution in Al-Ni-(Ti, Nb, Zr) powder blends by mechanical alloying." Materials Science and Engineering A 464(1-2): 306-314.

187. Sasaki, K. and H. S. Isaacs (2004). "Origins of Electrochemical Noise during Pitting Corrosion of Aluminum." Journal of The Electrochemical Society 151(3): B124-B133.

188. Sato, N. (1990). "An overview on the passivity of metals." Corrosion Science 31(0): 1-19.

189. Schmutz, P. and G. S. Frankel (1998). "Characterization of AA2024-T3 by scanning Kelvin probe force microscopy." Journal of The Electrochemical Society 145(7): 2285-2295.

190. Schneider, O., G. O. Ilevbare, et al. (2007). "In situ confocal laser scanning microscopy of AA2024-T3 corrosion metrology: III. Underfilm corrosion of epoxy-coated AA2024-T3." Journal of The Electrochemical Society 154(8): C397-C410.

191. Schneider, O., G. O. Ilevbare, et al. (2004). "In situ confocal laser scanning microscopy of AA 2024-T3 corrosion metrology II. Trench formation around particles." Journal of The Electrochemical Society 151(8): B465-B472.

192. Scholes, F. H., S. A. Furman, et al. (2006). "Chromate leaching from inhibited primers: Part I. Characterisation of leaching." Progress in Organic Coatings 56(1): 23-32.

193. Scholes, F. H., S. A. Furman, et al. (2006). "Corrosion in artificial defects. I: Development of corrosion." Corrosion Science 48(7): 1812-1826.

194. Scholes, F. H., A. E. Hughes, et al. (2009). "Interaction of Ce(dbp)(3) with surface of aluminium alloy 2024-T3 using macroscopic models of intermetallic phases." Corrosion Engineering Science and Technology 44(6): 416-424.

195. Scully, J. R., T. O. Knight, et al. (1993). "Electrochemical characteristics of the Al2Cu, Al3Ta and Al3Zr intermetallic phases and their relevancy to the localized corrosion of Al alloys." Corrosion Science 35(1–4): 185-195.

196. Searles, J. L., P. I. Gouma, et al. (2002). "Stress Corrosion Cracking of Sensitized AA5083 (Al-4.5Mg-1.0Mn)." Materials Science Forum 396-402: 1437-1442.

197. Semenov, A. M. (2001). "Effect of Mg additions and thermal treatment on corrosion properties of Al-Li-Cu-base alloys." Protection of Metals 37(2): 126-131.

198. Senkov, O., R. Bhat, et al. (2005). "Microstructure and properties of cast ingots of Al-Zn-Mg-Cu alloys modified with Sc and Zr." Metallurgical and Materials Transactions A 36(8): 2115-2126.

199. Seri, O. (1994). "The effect of NaCl concentration on the corrosion behavior of aluminum containing iron." Corrosion Science 36(10): 1789-1803.

200. Seri, O. and K. Tagashira (1986). "Effect of Manganese Content on Corrosion Characteristics of Al-Mn Alloys." Keikinzoku/Journal of Japan Institute of Light Metals 36(12): 806-812.

201. Sha, G. and A. Cerezo (2004). "Characterization of precipitates in an aged 7xxx series Al alloy." Surface and Interface Analysis 36(5-6): 564-568.

202. Shaw, B. A., G. D. Davis, et al. (1991). "The Influence of Tungsten Alloying Additions on the Passivity of Aluminum." Journal of The Electrochemical Society 138(11): 3288-3295.

203. Shaw, B. A., T. L. Fritz, et al. (1990). "The Influence of Tungsten on the Pitting of Aluminum Films." Journal of The Electrochemical Society 137(4): 1317-1318.

204. Sinko, J. (2001). "Challenges of chromate inhibitor pigments replacement in organic coatings." Progress in Organic Coatings 42(3–4): 267-282.

205. Song, R. G., W. Dietzel, et al. (2004). "Stress corrosion cracking and hydrogen embrittlement of an Al-Zn-Mg-Cu alloy." Acta Materialia 52(16): 4727-4743.

206. Souto, R. M., Y. González-García, et al. (2010). "Examination of organic coatings on metallic substrates by scanning electrochemical microscopy in feedback mode: Revealing the early stages of coating breakdown in corrosive environments." Corrosion Science 52(3): 748-753.

207. Speckert, L. and G. T. Burstein (2011). "Combined anodic/cathodic transient currents within nucleating pits on Al-Fe alloy surfaces." Corrosion Science 53(2): 534-539.

208. Stelling, O., A. Irretier, et al. (2006). New light-weight aluminum alloys with high Mg2Si-content by spray forming. 519-521: 1245-1250.

209. Sukiman, N. L., N. Birbilis, et al. (2010). Corrosion Maps For Aluminium Alloys. Proc. Conf. Corrosion And Prevention '10. Adelaide, Australia, Australasian Corrosion Association (ACA).

210. Suryanarayana, C., K. C. Rao, et al. (2008). "Preparation and characterization of microcapsules containing linseed oil and its use in self-healing coatings." Progress in Organic Coatings 63(1): 72-78.

211. Szklarska-Smialowska, Z. (1999). "Pitting corrosion of aluminum." Corrosion Science 41(9): 1743-1767.

212. Szklarska-Smialowska, Z. (2002). "Mechanism of pit nucleation by electrical breakdown of the passive film." Corrosion Science 44(5): 1143-1149.

213. Tahani, A., E. Chaieb, et al. (2003). "Electrochemical study of the influence of rolling on the resistance. The corrosion of alloy of aluminium 3003 in milieu 3% NaCl." Transactions of the SAEST (Society for Advancement of Electrochemical Science and Technology) 38(1): 43-46.

214. Tailleart, N., B. Gauthier, et al. (2009). Metallurgical and physical factors in improving the corrosion resistance of thermally sprayed semi-amorphous Al-Co-Ce coatings.

215. Tan, L. and T. R. Allen (2010). "Effect of thermomechanical treatment on the corrosion of AA5083." Corrosion Science 52(2): 548-554.

216. Taylor, S. R. and B. D. Chambers (2008). "Identification and Characterization of Nonchromate Corrosion Inhibitor Synergies Using High-Throughput Methods." Corrosion 64(3): 255-270.

217. Tedim, J., S. K. Poznyak, et al. (2010). "Enhancement of Active Corrosion Protection via Combination of Inhibitor-Loaded Nanocontainers." Acs Applied Materials & Interfaces 2(5): 1528-1535.

218. Teixeira, J. D. C., L. Bourgeois, et al. (2007). A study of the microstructural evolution and strengthening effects of non-spherical precipitates in an al-cu-based alloy. 561-565: 2317-2320.

219. Trueman, A. R. (2005). "Determining the probability of stable pit initiation on aluminium alloys using potentiostatic electrochemical measurements." Corrosion Science 47(9): 2240-2256.

220. Twite, R. L. and G. P. Bierwagen (1998). "Review of Alternatives to Chromate for Corrosion Protection of Aluminum Aerospace Alloys." Progress in Organic Coatings 33(2): 91-100.

221. Unocic, K. A., P. Kobe, et al. (2006). "Grain Boundary Precipitate Modification for Improved Intergranular Corrosion Resistance." Materials Science Forum 519-521: 327-332.

222. Usta, M., M. E. Glicksman, et al. (2004). "The effect of heat treatment on Mg2Si coarsening in aluminum 6105 alloy." Metallurgical and Materials Transactions A: Physical Metallurgy and Materials Science 35 A(2): 435-438.

223. Vetrano, J. S., C. H. Henager Jr, et al. (1998). Use of Sc, Zr and Mn for grain size control in Al-Mg alloys.

224. Vetrano, J. S., R. E. Williford, et al. (1997). Influence of microstructure and thermal history on the corrosion susceptibility of AA5083. TMS Annual Meeting, Orlando, FL, USA.

225. Voevodin, N. N., V. N. Balbyshev, et al. (2003). "Nanostructured coatings approach for corrosion protection." Progress in Organic Coatings 47(3–4): 416-423.

226. Walmsley, J. C., Ø. Sævik, et al. (2007). "Nature of segregated lead on electrochemically active AlPb model alloy." Journal of The Electrochemical Society 154(1): C28-C35.

227. Wang, D., D. Battocchi, et al. (2010). "In situ monitoring of a Mg-rich primer beneath a topcoat exposed to Prohesion conditions." Corrosion Science 52(2): 441-448.

228. Wang, W. t., X. m. Zhang, et al. (2010). "Influences of Ce addition on the microstructures and mechanical properties of 2519A aluminum alloy plate." Journal of Alloys and Compounds 491(1-2): 366-371.

229. Wei, R. P. (2001). "A model for particle-induced pit growth in aluminum alloys." Scripta Materialia 44(11): 2647-2652.

230. Westermann, I., O. S. Hopperstad, et al. (2012). "Effect of alloying elements on stage-III work-hardening behaviour of Al -Zn-Mg(-Cu) alloys." International Journal of Materials Research 103(5): 603-608.

231. White, P. A., A. E. Hughes, et al. (2009). "High-throughput channel arrays for inhibitor testing: Proof of concept for AA2024-T3." Corrosion Science 51(10): 2279-2290.

232. White, S. R., N. R. Sottos, et al. (2001). "Autonomic healing of polymer composites." Nature 409(6822): 794-797.

233. Williams, D. E., J. Stewart, et al. (1994). "The nucleation, growth and stability of micropits in stainless steel." Corrosion Science 36(7): 1213-1235.

234. Williams, G. and H. N. McMurray (2003). "Anion-exchange inhibition of filiform corrosion on organic coated AA2024-T3 aluminum alloy by hydrotalcite-like pigments." Electrochemical and Solid State Letters 6(3): B9-B11.

235. Wu, B., J. R. Scully, et al. (1997). "Cooperative stochastic behavior in localized corrosion: I. Model." Journal of The Electrochemical Society 144(5): 1614-1620.

236. Wu, D. Y., S. Meure, et al. (2008). "Self-healing polymeric materials: A review of recent developments." Progress in Polymer Science 33(5): 479-522.

237. Xu, D. K., N. Birbilis, et al. (2011). "Effect of solution treatment on the corrosion behaviour of aluminium alloy AA7150: Optimisation for corrosion resistance." Corrosion Science 53(1): 217-225.

238. Yang, B. J., J. H. Yao, et al. (2009). "Al-rich bulk metallic glasses with plasticity and ultrahigh specific strength." Scripta Materialia 61(4): 423-426.

239. Yasakau, K. A., M. L. Zheludkevich, et al. (2007). "Role of intermetallic phases in localized corrosion of AA5083." Electrochimica Acta 52(27 SPEC. ISS.): 7651-7659.

240. Yin, T., M. Z. Rong, et al. (2007). "Self-healing epoxy composites - Preparation and effect of the healant consisting of microencapsulated epoxy and latent curing agent." Composites Science and Technology 67(2): 201-212.

241. Yu, Y., Ø. Sævik, et al. (2005). "Effect of annealing temperature on anodic activation of rolled AA8006 aluminum alloy by trace element lead." Journal of The Electrochemical Society 152(9): B327-B333.

242. Yu, Y., Ø. Saevlk, et al. (2004). "Characterization of lead enrichment on electrochemically activated binary Al-Pb model alloy." Materials Forum 28: 270-276.

243. Zamin, M. (1981). "ROLE OF Mn In the Corrosion Behavior of Al-Mn Alloys." Corrosion 37(11): 627-632.

244. Zawrah, M. and L. Shaw (2003). "Microstructure and hardness of nanostructured Al–Fe–Cr–Ti alloys through mechanical alloying." Materials Science and Engineering: A 355(1–2): 37-49.

245. Zeng, F.-l., Z.-l. Wei, et al. (2011). "Corrosion mechanism associated with Mg2Si and Si particles in Al–Mg–Si alloys." Transactions of Nonferrous Metals Society of China 21(12): 2559-2567.

246. Zhang, W. and G. S. Frankel (2003). "Transitions between pitting and intergranular corrosion in AA2024." Electrochimica Acta 48(9): 1193-1210.

247. Zhao, Y. H., X. Z. Liao, et al. (2004). "Microstructures and mechanical properties of ultrafine grained 7075 Al alloy processed by ECAP and their evolutions during annealing." Acta Materialia 52(15): 4589-4599.

248. Zheludkevich, M. L., I. M. Salvado, et al. (2005). "Sol-gel coatings for corrosion protection of metals." Journal of Materials Chemistry 15(48): 5099-5111.

249. Zhong, J., K. Feng, et al. (2011). Effect of RE on the mechanical properties of 7075 Al alloy. 150-151: 1286-1289.

250. Zhou, X., N. Birbilis, et al. (2010). "Kinetic Stability of Aluminium." Corrosion & Prevention Corrosion, Adelaide, Australia.

251. Zhou, X., Y. Liu, et al. (2011). "Near-Surface Deformed Layers on Rolled Aluminum Alloys." Metallurgical and Materials Transactions A 42(5): 1373-1385.

252. Zhou, X., C. Luo, et al. (2012). "Study of localized corrosion in AA2024 aluminium alloy using electron tomography." Corrosion Science 58: 299-306.

253. Zieliński, A., J. Chrzanowski, et al. (2004). "Influence of retrogression and reaging on microstructure, mechanical properties and susceptibility to stress corrosion cracking of an Al-Zn-Mg alloy." Materials and Corrosion 55(2): 77-87.

254. Zin, I. M., R. L. Howard, et al. (1998). "The mode of action of chromate inhibitor in epoxy primer on galvanized steel." Progress in Organic Coatings 33(3–4): 203-210.

255. Zou, L., Q.-l. Pan, et al. (2007). "Effect of minor Sc and Zr addition on microstructures and mechanical properties of Al-Zn-Mg-Cu alloys." Transactions of Nonferrous Metals Society of China 17(2): 340-345.

CHAPTER 2

Equal-Channel Angular Pressing and Creep in Ultrafine-Grained Aluminium and Its Alloys

Vaclav Sklenicka[1, 2], Jiri Dvorak[1, 2], Milan Svoboda[1, 2], Petr Kral[1] and Marie Kvapilova[1]

[1] Institute of Physics of Materials Academy of Sciences of the Czech Republic, Brno, Czech Republic

[2] CEITEC-IPM, Institute of Physics of, Materials, Academy of Sciences of the Czech Republic, Brno, Czech Republic

1. INTRODUCTION

Creep strength and ductility are the key creep properties of creep-resistant materials but these properties typically have opposing characteristics. Thus, materials with conventional grain sizes may be strong or ductile but there are rarely both. In this connection, recent findings of high strength and good ductility in several submicrometer metals and alloys are of special interest [1]. Reduction of the grain size of a polycrystalline material can be successfully produced through advanced synthesis processes such as the electrodeposition technique [2] and severe plastic deformation SPD [1,3-6]. Although creep is an exceptionally old area of research, above mentioned processing techniques have become available over the last two decades which provide an opportunity to expand the creep behaviour into new areas that were not feasible in earlier experiments. Creep testing of nanocrystalline (grain size $d < 100$ nm) and ultrafine-grained ($d < 1$ μm) materials is characterized by features that may be different from those documented for coarse-grained materials and thus cannot easily be compared.

Processing through the application of severe plastic deformation (SPD) is now an accepted procedure for producing bulk ultrafine-grained materials having grain sizes in the submicrometer or nanometer range. The use of SPD enhances certain material properties through the introduction of an ultrafine-grained microstructure. The ultrafine size of the grains in the bulk materials generally leads to significantly improved properties by comparison with polycrystalline materials having conventional grain sizes of the same chemical composition. Several SPD processing techniques are currently available but the most attractive technique is equal-channel angular pressing (ECAP), where the sample is pressed through a die constrained within a channel bent through an abrupt angle [4]. There are numerous reports of the processing of various pure metals and metallic alloys by ECAP and many of these reports involve a characterization of the microstructure and an investigation of the mechanical properties at ambient temperatures. There are also several reports of the tensile properties of the as-pressed materials at elevated temperatures with a special emphasis on the potential for achieving high superplastic elongations. However, the tests at elevated temperatures are invariably conducted under conditions of constant strain rate and, by contrast, only very limited reports are available describing the creep behaviour of aluminium and some aluminium alloys. Furthermore, the results for high-purity aluminium, which are the most extensive available to date, appear anomalous because under some testing conditions of stress and temperature the measured minimum or steady-state creep rates in the pressed materials with ultrafine grain sizes where slower than in the same material in a coarse-grained unpressed condition.

This chapter was initiated to provide basic information on the creep behaviour and microstructural characteristics of aluminium and some aluminium alloys. The chapter has arisen in connection with long-term research activity of the Advanced High Temperature Materials Group at the Institute of Physics of Materials, Academy of Sciences of the Czech Republic in Brno, Czech Republic. Thus, the objective of this chapter is to present an overview of some results of our current research in creep behaviour and a link between the microstructure and the creep properties of ultrafine-grained aluminium based alloys. Throughout the text, our results are compared with theoretical models and relevant experimental observations published in the literature.

2. THE DEVELOPMENT OF PROCESSING USING EQUAL-CHANNEL ANGULAR PRESSING (ECAP)

Processing by severe plastic deformation (SPD) may be defined as those metal forming procedures in which a very high strain is imposed on a bulk solid without the introduction of any significant change in the overall dimensions of the solid and leading to the production of exceptional grain refinement to that the processed bulk solids have 1000 or more grains in section [4]. Of a wide diversity of new SPD procedures, equal-channel angular pressing (ECAP) is an especially attractive processing technique. It is relatively simple procedure which can be applied to fairly large billets of many materials ranging from pure metals to precipitation-hardened alloys, intermetallics and metal-matrix composites.

2.1. Principles Of Ecap

The principle of ECAP is illustrated schematically in Figure 1. For the die shown in Figure 1, the internal channel is bent through an abrupt angle, Φ, and there is an additional angle, Ψ, which represents outer arc of curvature where the two channels intersect. The sample, in the form of a rod or bar, is machined to fit within channel and the die is placed in some form of fuss so that the sample can be pressed through the die using a plunger. The nature of the imposed deformation is simple shear which occurs as the billet passes through the die. The retention of the same cross-sectional area when processing by ECAP, despite the introduction of very large strains, is the important characteristic of SPD processing and it is characteristic which distinguishes this type of processing from conventional metal-working operations such as rolling, extrusion and drawing. Since the cross-sectional area remains unchanged, the same billet may be pressed repetitively to attain exceptionally high strain.

Aluminium and its alloys used in this investigation were pressed using an experimental facility for ECAP installed in the Institute of Physics of Materials, Academy of Sciences of the Czech Republic (Figure 2). The die was placed on a testing machine Zwick. ECAP was conducted mostly at room temperature with a die that had internal angle 90° between two parts of the channel and an outer arc of curvature of ~ 20°, where these two parts intersect. It can be shown from first principles that these angles lead to an imposed strain of ~ 1 in each passage of the sample. The ECAP die involved the use of billets of the length of ~ 50 – 60 mm with square cross-section of 10 mm x 10 mm. The velocity of plunger was 10 mm/min.

2.2. The Processing Routes In Ecap

The use of repetitive pressing provides an opportunity to invoke different slip systems on each consecutive pass by simply rotating the samples in different ways. The four different processing routes are summarized schematically in Figure 3 [7]. In route A the sample is pressed without rotation, in route B_A the sample is rotated by 90° in alternate directions between consecutive passes, in route $B_C$the sample is rotated by 90° in the same sense (either clockwise or counter clockwise) between each pass and in route C the sample is rotated by 180° between passes. The distinction between these routes and the difference in number of ECAP passes may lead to variations both in the macroscopic distortions of the individual grains [8] and in the capability to develop a reasonably homogeneous and equiaxed ultrafine-grained microstructure.

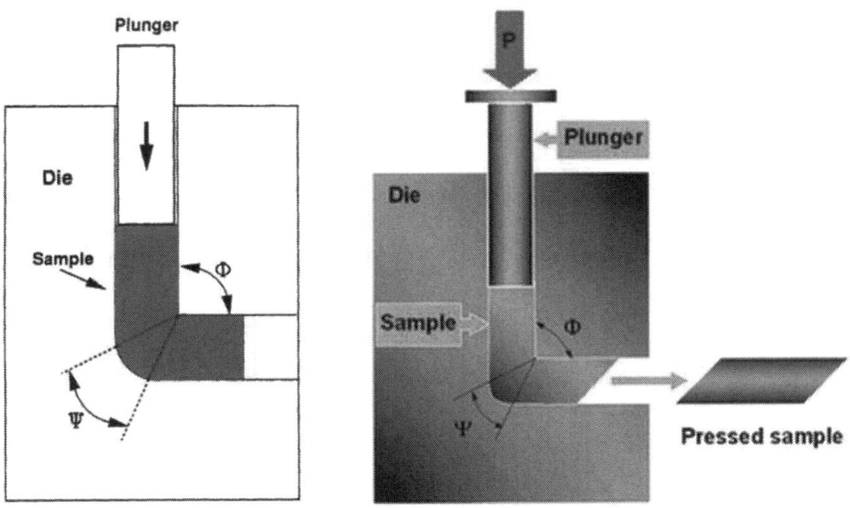

Figure 1. Principle of ECAP.

In this work the ECAP pressing was conducted in such a way that one or repetitive pressing was conducted followed either route A, B (route B_C was used only) or C. Detailed examinations of the effect of different processing routes showed that route B_C leads to the most rapid evolution into an array of high-angle grain boundaries [9,10]. The result is explained by considering the shearing patterns developed in the samples during each processing route. Thus, the route B_C is most probably the optimum ECAP processing route at least for the pressing of pure aluminium and its alloys [4].

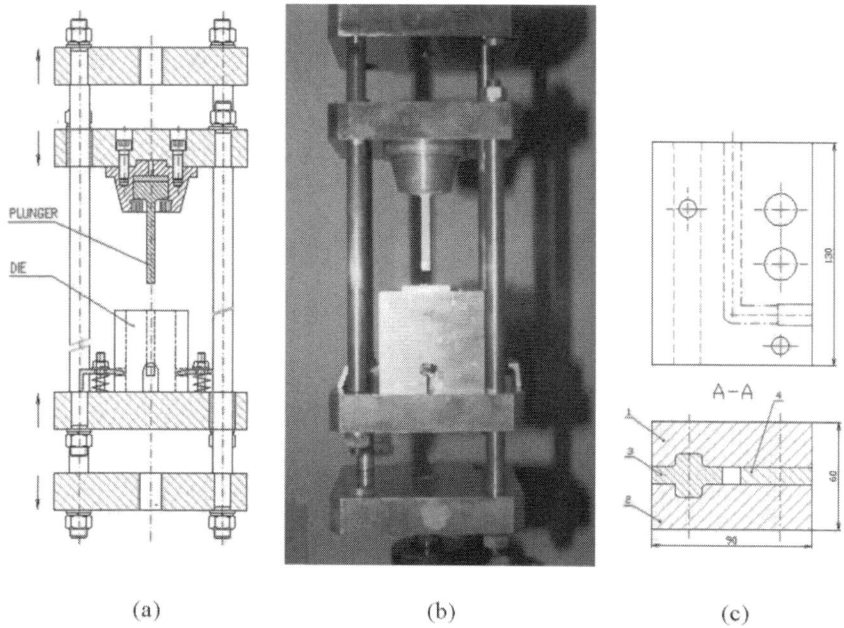

Figure 2. Adaptation of testing ZWICK machine for ECAP pressing (a, b), and (c) sketch of ECAP die design.

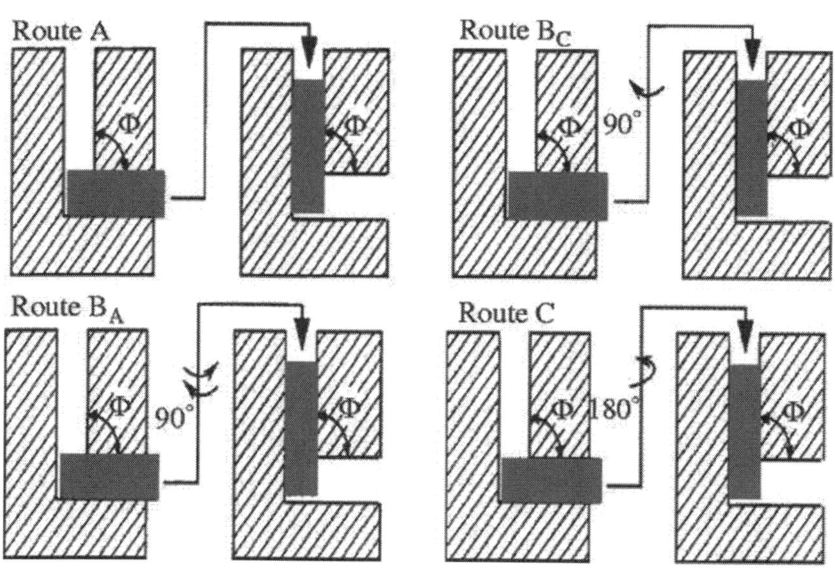

Figure 3. Schematic of four ECAP routes for repetitive pressing.

2.3. Mechanical Properties And Defects Achieved Using Ecap

During the last two decades it has been demonstrated that an ultrafine-grained structure of materials processed by ECAP may lead to significantly higher strength and hardness but to a reduction in the ductility [4]. In this connection after ECAP the mechanical properties were tested mostly at room temperature using a testing machine operating at a constant rate of 2.0 x $10^{-4}s^{-1}$ of crosshead displacement.

2.3.1. Tensile Properties

Tensile tests were conducted at 293 K on pure aluminium after processing by ECAP for samples after different number of ECAP passes. In limited extent mechanical tests were performed on the samples after ECAP and static annealing at 473 K [11]. In Figure 4 the tensile data are summarized as a function of the number of passes. It is apparent from these figures that a very significant increase in yield and ultimate tensile stress occurred after the first pressing. The subsequent pressing further increased yield and ultimate stress values but to a lower rate. Further, a saturation of the level of both the parameters was attained after four passes.

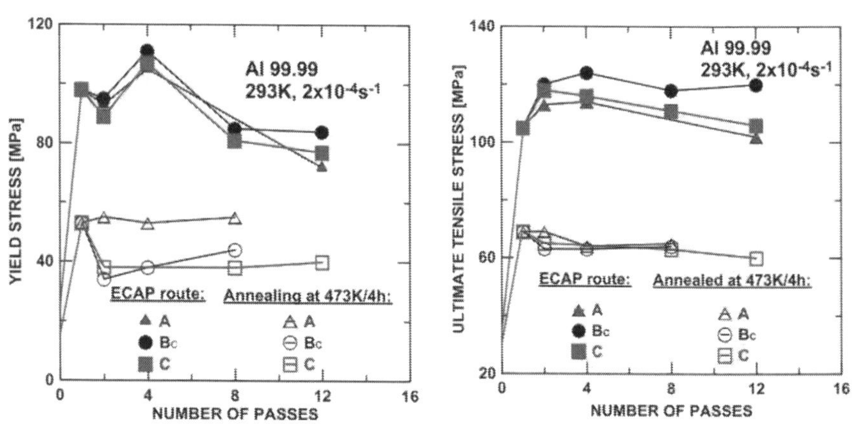

Figure 4. Influence of different ECAP routes and different number of ECAP passes on (a) yield stress, and (b) ultimate tensile stress after static annealing.

From Figure 4 it can be also noticed that static annealing at 473 K leads to a substantial decrease in the level of yield and ultimate tensile stress values due to diffusion based recovery processes for all the ECAP processed samples. No significant differences in mechanical properties among the ECAP process routes examined were found. Further, from Figure 4 is clear that although the levels of the tensile data for ECAPed Al highly decrease with the number of ECAP passes, the stress levels after 8 passes are much higher than the stress levels in the annealing state and these differences come to more than twice. This result indicates that, when compared with the tensile behaviour of the annealed state, the flow stress is considerably improved through the application of ECAP [11,12].

2.3.2. Hardness Measurements

Figure 5a shows Vickers microhardness plotted against the number of ECAP passes for extremely high purity aluminium (99.99%) [12]. The hardness increases up to two passes to take a maximum due to the very high dislocation density. However, subsequent passes lead to a decrease in the hardness because many of the subgrain boundaries evolve into high-angle grain boundaries. Figure 5b shows Vickers microhardness plotted against different periods of time of a static annealing at 473 K for pure (99.99%) Al processed by ECAP by two different processing routes. A pronounced decrease of microhardness with an increase of annealing time can be explained by significant grain growth and softening of pressed material during an annealing exposures [11].

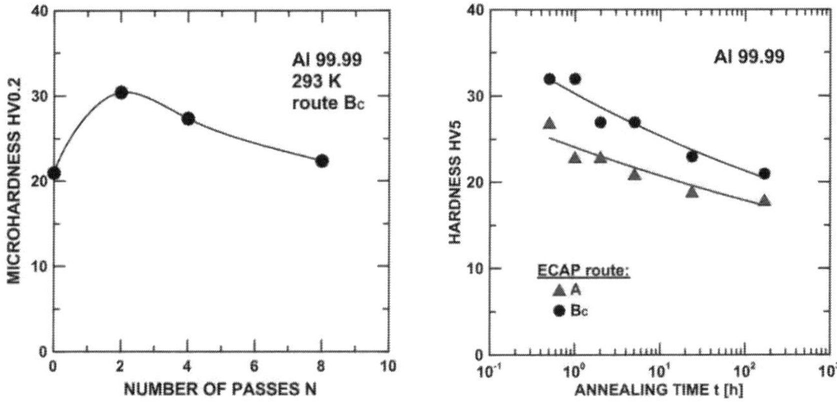

Figure 5. Hardness changes (a) with respect to number of ECAP passes, and (b) as a function of annealing time at 473 K for two different ECAP routes.

2.3.3. Nanoporosity After Ecap Processing

It is generally recognized that the ECAP process could produce a submicrocrystalline bulk material with a relatively uniform structure and 100% density for a wide range of materials from pure metals, solid-solution alloys, commercial alloys, to metal matrix composites [1]. However, the previously performed analysis of the data on the influence of the number of passes of equal-channel angular pressing on the elastic-plastic properties and defect structure of pure aluminium demonstrated that these characteristics of mechanical properties are substantially affected by the evolution of the nanoporosity formed during equal-channel angular pressing [13-15]. Thus, to determine the total volume of nanoporosity which could be generated by ECAP, two selected samples of pure aluminium were pressed for a total of one (specimen A1) and four (specimen A4) ECAP passes, respectively, and for comparison reasons some part of these specimens were underwent by subsequent pressurization treatment by high hydrostatic pressure [16]. The samples were investigated by small-angle X-ray scattering (SAXS) and dilatometry [13].

Some differences were found in the fractional volume of the nanopores $\Delta V/V$ when compared specimen A1 to specimen A4. The values $\Delta V/V_{max}$ correspond to the as-pressed state of specimens (after ECAP only) and the values $\Delta V/V_{min}$ were evaluated for the state after ECAP and subsequent pressurization which represents a rejuvenative treatment for elimination of nanopores. The evaluated values are $\Delta V/V_{max} = 5x10^{-3}$ and $\Delta V/V_{min} = 2.5.10^{-3}$ for specimen A1 and $\Delta V/V_{max} = 7x10^{-3}$ and $\Delta V/V_{min} = 3x10^{-3}$ for specimen A4, respectively. No substantial difference in the average size of the nanopores (~ 20-30nm) was found between the specimens investigated. The values $\Delta V/V$ determined by small-angle X-ray scattering and dilatometry were about the same; e.g. the fractional volume $\Delta V/V_{min} = 3x10^{-3}$ by (SAXS) of specimen A4 agreed very well with $\Delta V/V_{min} = 2.5x10^{-3}$ as determined by dilatometry. On the basis of the aforementioned results we can conclude that ECAP deformation achieves strongly enhanced concentration of vacancy agglomerates type defects. The effect of the spectrum of the point defects and the internal stresses on elasticity and anelasticity of ECAPed aluminium has been reported elsewhere [17].

In recent years using a back-pressure ECAP facilities [4] has become an area of special interest. An important advantage in imposing a back-pressure may be a decrease of nanoporosity in the pressed material [18]. However, additional experiments are needed to evaluate the role of a back-pressure in elimination of nanoporosity.

3. MICROSTRUCTURAL FEATURES OF ULTRAFINE-GRAINED MATERIALS

Ultrafine-grained (UFG) materials processed by ECAP differ qualitatively and quantitatively from their coarse-grained (CG) counterparts in terms of their characteristic structural parameters and thus their creep behaviour cannot be easily compared with that documented for CG materials. It is important to note in this respect that UFG materials are characterized by great extension of internal interfaces; therefore, grain boundary diffusion processes have to be involved in the formation of their structure-sensitive properties, especially at elevated temperature [19].

The characteristics of the microstructures introduced by ECAP have been evaluated in numerous investigations [4]. However, most of these earlier investigations employed transmission electron microscopy (TEM) for determinations of the grain sizes produced by ECAP and the nature of any dislocation interactions occurring within grains. The application of modern imaging methods to the examination of microstructures in UFG materials processed by ECAP has permitted a more detailed investigation of a possible link between internal microstructures of UFG metals and alloys and their mechanical and/or creep behaviour [4]. Diffraction-based techniques for localized crystal orientation measurements, such as electron backscatter diffraction (EBSD), are of central importance today for characterizing fine-scale microstructural features [20-23].

The new experimental technique of EBSD considerably extended the possibilities of metallography to estimate reliably the quantitative structural characteristics of materials [23]. It enables the numerical classification of boundaries separating the regions of different orientations of their lattice structure. The magnitude of the mutual misorientation can be continuously selected and thus the regions with a misorientation less than a prescribed value as well as their boundaries can be recognized. There is a vast literature devoted to the observation by EBSD and precisely defined misorientation of boundaries and the conventional grain boundary classification based on suitably polished and etched planar surfaces as observed by optical microscopy or by boundaries observed by electron microscopy and EBSD (see e.g. [24]). As can be expected, the EBSD method is more reproducible, independent of detailed etching conditions etc., and the surface area intensities are usually higher (equivalently, the mean random profile chord is smaller). In this section a division of boundaries into true subboundaries with misorientations $\Delta < 10°$, transitional subboundaries with $10° \leq \Delta < 15°$ and high-angle grain boundaries with $\Delta \geq 15°$ was made.

Such an approach is of primary importance in the examination of materials produced by severe plastic deformation (SPD), without change of shape, producing materials with ultrafine grains (e.g. [3,5]) and considerably different properties in comparison with CG materials. The reason for this difference is to a certain degree purely geometric and consists in different grain and subgrain boundary structures, which play an important role in mechanical, thermal and other properties.

This section describes the results of structural examinations of high purity aluminium and its selected precipitation-strengthened alloys processed by ECAP. The microstructure was revealed by TEM, SEM and EBSD and analyzed quantitatively by stereological methods. The various factors influencing the as-pressed microstructures including the total strain imposed in ECAP processing, the processing routes and the nature of materials are examined in detail.

3.1. Experimental Materials and Their Microstructure After Ecap

3.1.1. Pure Aluminium

The aluminium used in this investigation was an extremely coarse-grained (grain size ~ 5 mm) high purity (99.99%) Al supplied in the form of rods. The rods were cut into short billets having a length of ~ 60 mm and a cross-section 10 mm x 10 mm. ECAP was conducted at room temperature using route A, B_C and C. Full details on the processing have been described elsewhere [25-27].

TEM results have shown that one ECAP pass leads to a substantial reduction in the grain size (~ 1.4 μm), and the microstructure consists of parallel bands of grains oriented in the shearing direction. The microstructure is very inhomogeneous and the grain size varies from location to location. The inhomogeneous nature of the microstructure may reflect the coarse grain size (~ 5 mm) prior to ECAP. The grains subsequently evolve upon subsequent ECAP passes into a reasonably equiaxed and homogeneous microstructure with an average grain size of ~1 μm regardless of the particular ECAP routes. The microstructure is essentially homogeneous after four ECAP passes, although a tendency for grain elongation in the direction of the shear direction of the last pressing operation is retained. Figure 6 gives an example of the microstructure in the cross-section normal to the pressing direction after four subsequent ECAP passes performed in different routes. TEM micrographs in Figure 7 give an example of the microstructure in cross-section after four and eight subsequent ECAP passes by route B_c and C, respectively. The EBSD grain maps in Figure 8 indicate little dependence of the grain boundary disorientation distribution on the ECAPed Al processed by route B_c.

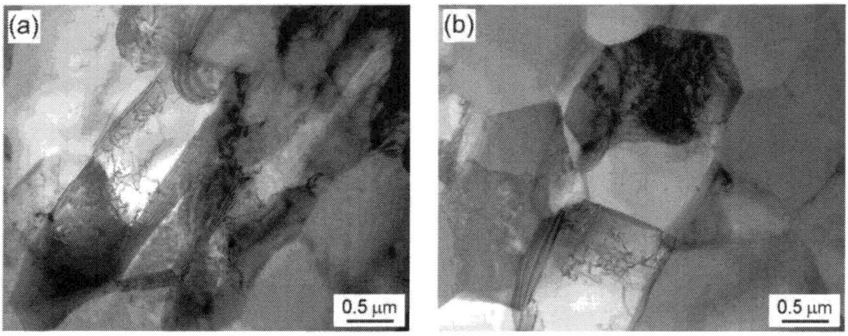

Figure 6. TEM micrographs of aluminium after four subsequent ECAP passes on route (a) A, and (b) B.

Figure 7. Typical microstructures and associated SAED patterns after passage through the die for (a) 4 pressings, route B and (b) 8 pressings, route C.

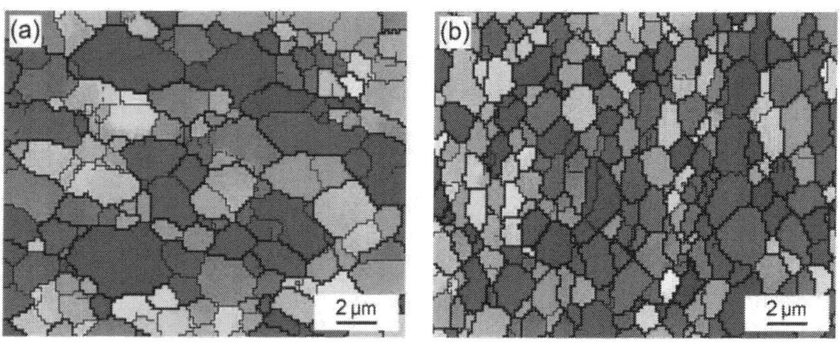

Figure 8. Grain maps for ECAPed Al after: (a) 4 passes, and (b) 8 passes by route B (EBSD).

It can be expected that the creep behaviour of the ultrafine-grained pure aluminium will critically depend on the thermal stability of the microstructure. To explore the thermal stability of ECAP processed aluminium load-less annealing was conducted at temperature of 473 K for different periods of time (i.e. at the temperature of the intended creep tests). Microscopic examination revealed that the post-ECAP annealing makes the ECAP microstructure quite unstable and a noticeable grain growth occurs at the very beginning of annealing (Table 1). Simultaneously, annealing at 473 K gives measurable change in the Vickers microhardness.

Table 1. Thermal stability and Vickers microhardness of the ECAP aluminium.

Annealing conditions	ECAP 4 passes route A		ECAP 4 passes route B	
	grain size [μm]	microhardness HV5	grain size [μm]	microhardness HV5
no annealing	0.9	37	0.9	38
473 K/ 0.5 h	6.6	27	4.5	32
473 K/ 1 h	7.9	23	4.8	32
473 K/ 2 h	7.3	23	4.8	27
473 K/ 5 h	7.3	21	5.3	27
473 K/ 24 h	12.2	19	5.0	23
473 K/ 168h	13.4	18	10.4	21

3.1.2. Precipitation-Strengthened Aluminium Alloys

In evaluating the microstructure characteristics of ultrafine-grained materials processed by ECAP at elevated and high temperatures, it is very important to recognize that these ultrafine-grained microstructures are frequently unstable at these temperatures as it was just demonstrated by the above mentioned results of thermal instability of pressed pure aluminium. However, it is often feasible to retain an array of ultrafine grains even at very high temperatures by using materials containing second phases or arrays of precipitates. This was a reason why two precipitation-strengthened aluminium alloys were used in this investigation.

It has been shown that addition to aluminium alloys of even very small amounts of Sc (typically, ~ 0.2wt.%) strongly improves the microstructures of the alloys and their mechanical properties so that these alloys are suitable for use in engineering applications [28]. Scandium additions of ~ 0.2wt.%Sc to pure aluminium are sufficient to more or less retain a small grain size at elevated temperatures [29]. Further, some reports have demonstrated that it is possible to achieve high ductilities in Al-Mg-Sc alloys by using ECAP to introduce an exceptionally small grain size [30]. The creep behaviour of conventional Al-Mg alloys is extensively described in the literature. The synergy of solid-solution strengthening and precipitate strengthening has, however, not been extensively studied at elevated and high temperatures [31]. Very little information is available at present on the creep properties of ultrafine-grained Al-Sc and Al-Mg-Sc alloys [32-38]. Accordingly, the present investigation was initiated to provide a more complex information on the creep behaviour of these aluminium alloys in their ultrafine-grained states.

An Al-0.2wt.%Sc alloy was produced by diluting an Al-2.0wt.%Sc master alloy with 99.99wt.% pure aluminium. The resulting ingots were subjected to a homogenization and grain-coarsening treatment at 893 K for 12 hours and then aged in air at 623 K for 1 hour. In the as-fabricated condition, the extremely coarse grain size was measured as ~ 8 mm. The ECAP was conducted at the Institute of Physics of Materials AS CR Brno, Czech Republic, using the same die and procedure as it was reported earlier for pure aluminium (i.e. up to a total 8 ECAP passes at room temperature). The details concerning an Al-0.2wt.%Sc alloy have been reported elsewhere [33-35]. The ternary Al-Mg-Sc alloy was fabricated at the Department of Materials Science and Engineering, Faculty of Engineering, Kyushu University, Fukuoka, Japan. The alloy contained 3wt.%Mg and 0.2wt.%Sc and it was prepared from 99.99% purity Al, 99.999% purity Sc and 99.9% purity Mg. Full details on the fabrication procedure are given elsewhere [32] but, briefly, the alloy was cast, homogenized in air for 24 h at 753 K and solution treated for 1 h at 883 K. In the as-fabricated condition, the grain size was about 200 μm. Again, the ECAP was conducted using a solid die that had 90° angle between the die channels and each sample was pressed at room temperature repetitively for a total of eight passes by route B_C.

Figures 9a,b and 10a,b show the microstructure of Al-0.2wt.%Sc and Al-3wt.%Mg-0.2wt.%Sc alloys in their as-pressed states. Experiments on Al-0.2wt.%Sc and Al-3wt.%Mg-0.2wt.%Sc alloys revealed that processing by ECAP reduced the grain size to ~ 0.4 μm and subsequent annealing at 623 K and 1 h and creep testing gave the grain sizes ~ 0.9 μm for an Al-0.2wt.%Sc alloy and ~ 1.5 μm for an Al-3wt.%Mg-0.2wt.%Sc alloy, respectively. Figures 9c,d and 10c,d give examples of the microstructure of the alloys in the longitudinal sections parallel to the pressing direction after creep exposures at 473 K. As will be shown later on no substantial difference in the relative fractions of high-angle ($\theta > 15°$) grain boundary population after ECAP was found between the alloys investigated. These fractions were slightly increased during creep exposure up to an average value ~ 70%.

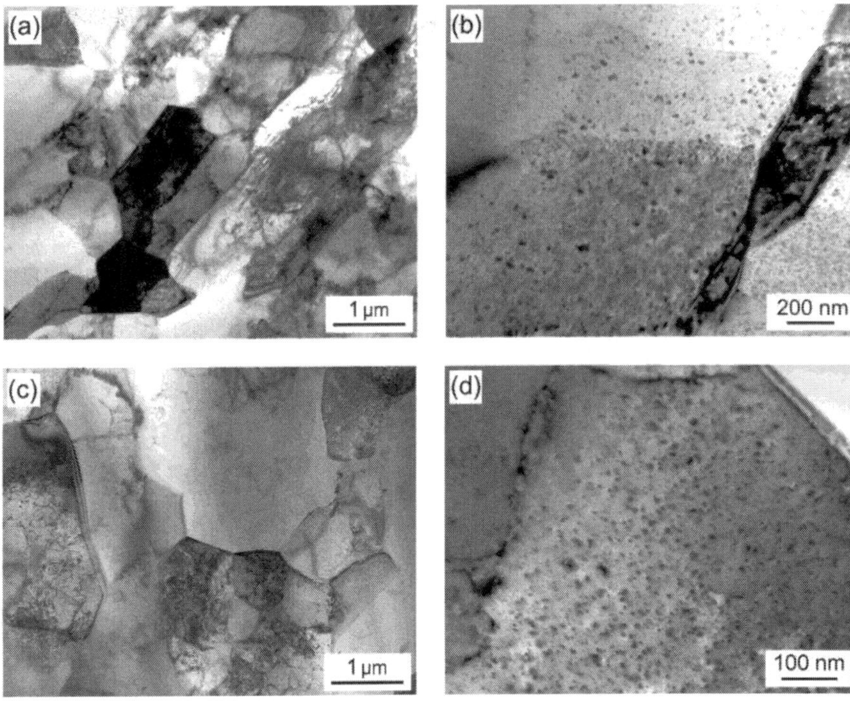

Figure 9. Microstructure in the Al-0.2wt.%Sc alloy: (a) and (b) after ECAP (B_C, 8 passes) and annealing for 1 h at 623 K, (c) and (d) after creep at 473 K.

Figures 9b,d and 10b,d exhibit TEM micrographs, which demonstrate the presence of coherent Al_3Sc precipitates within the matrices of both alloys. Al_3Sc precipitates are indicated by a coherency strain contrast [39]. A mean size of precipitates was ~ 5 nm after creep testing of the Al-3wt.%Mg-0.2wt.%Sc alloy and a mean size of precipitates slightly large was recorded for the Al-0.2wt.%Sc alloy (~ 6 nm [35]). Figure 10d shows dislocation microstructure observed after creep exposure of ECAPed Al-3wt.%Mg-0.2wt.%Sc. The dislocation pairs present in both alloys containing the smallest precipitate radii are very frequent. For larger precipitates the dislocations are pinned efficiently by Al_3Sc precipitates as climbing becomes slower. Figure 11a exhibits TEM micrograph of an Al-3wt.%Mg-0.2 wt.%Sc alloy showing the presence of coherent Al_3Sc precipitates in an unpressed sample, and Figure 11b presents precipitate size distribution in an unpressed alloy.

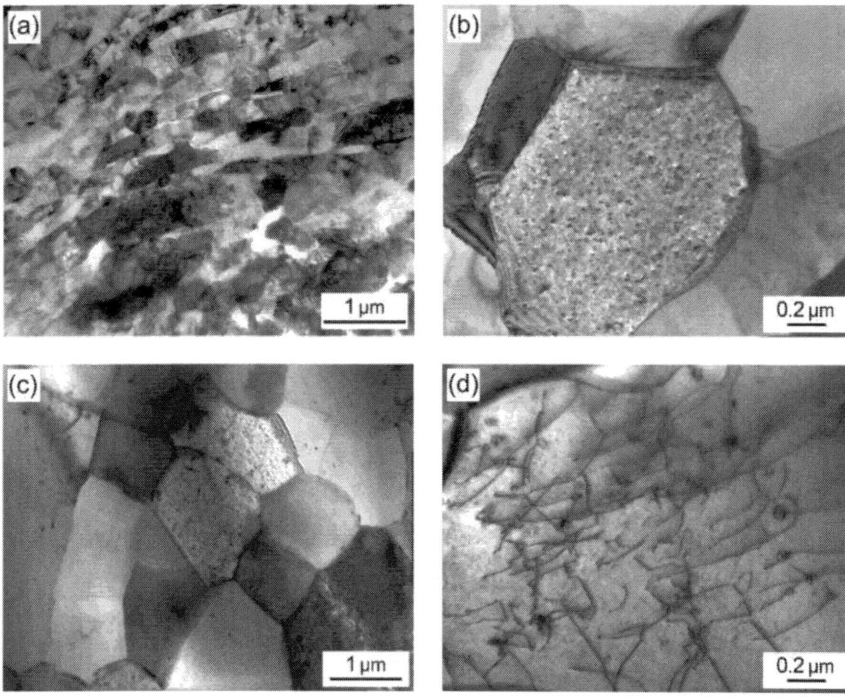

Figure 10. Microstructure in the Al-3wt.%Mg-0.2wt.%Sc alloy: (a) and (b) after ECAP (B_C, 8 passes) and annealing for 1 h at 623 K, (c) and (d) after creep at 473 K.

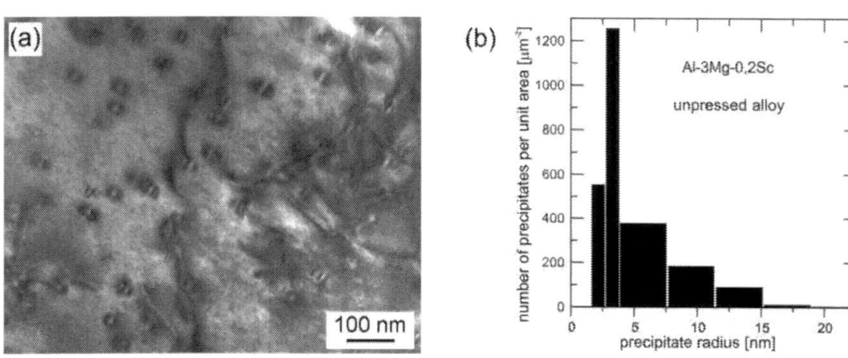

Figure 11. TEM micrograph showing the presence of coherent Al_3Sc precipitates in unpressed sample, and (b) precipitate size distribution in unpressed sample.

3.2. Microstructure Developed During Creep

3.2.1. Pure Aluminium

It can be expected that the creep behaviour of the UFG material will be influenced critically upon the subsequent thermal stability of its microstructure. To explore this effect microscopic examination of grain size change in pure aluminium during creep exposure at 473 K and 15 MPa were performed. It is important to note that each creep specimen was heated to the testing temperature in the furnace of the creep testing machine over a period of ~ 2h and then held at the testing temperature for further ~ 2h in order to reach thermal equilibrium. Consequently, the microstructure characteristics of the ECAP material at the onset of the creep testing were similar to that shown in Table 1. No substantial coarsening of grains has been observed during creep exposure at 473 K (see Table 2).

Table 2. Grain size of the ECAP material after creep at 473 K and 15 MPa.

Specimen	ECAP conditions	Grain size [μm]	Time to fracture [h]
A4	route A, 4 passes	6.4	79
A8	route A, 8 passes	7.0	26
A12	route A, 12 passes	6.7	17
B4	route B, 4 passes	8.7	62
B8	route B, 8 passes	7.2	60
B12	route B, 12 passes	8.8	39

TEM observations were used also to established details of microstructure evolution during creep. The micrographs in Figure 12a,b illustrate a dislocation substructure inside the grains. The dislocation lines were wavy and occasionally tangled with each other. It is know that large grains in UFG materials contain dislocations while grains smaller than a certain size are dislocation free [3,6]. EBSD measurements were taken to determine the grain boundary misorientation and the value of relative fraction of a high-angle grain boundary ($\theta > 15°$) population (for details see 3.3.2.).

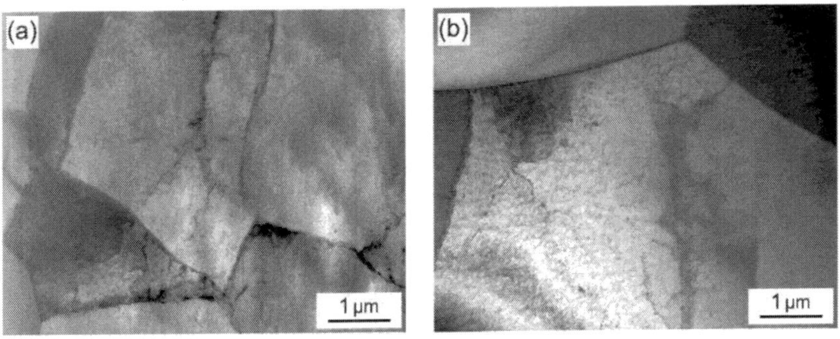

Figure 12. TEM micrographs from the longitudinal section of an aluminium processed by ECAP route B_c (a) after 1 ECAP pass and creep, (b) after 8 ECAP passes and creep. Creep at 473 K and 15 MPa.

3.2.2. Precipitation-Strengthened Alloys

For comparison reasons, some results of microstructural changes in Al-0.2wt.%Sc and Al-3wt.%Mg-0.2wt.%Sc alloys during creep were presented earlier in 3.1.2. It was found that creep exposures of an Al-0.2wt.%Sc alloy at 473 K and 20 MPa caused the changes in (sub)grain sizes initially resulting from ECAP pressing. Figure 13a shows the microstructure after 8 ECAP passes and subsequent creep exposure. TEM analysis revealed that the average (sub)grain size increases from ~ 0.4 um to ~ 1.3 µm after creep exposure. The (sub)grain growth was effected by presence of coherent Al_3Sc precipitates (Figure 13b) which to some extent pinned the boundaries against their migration and restricted the movement of dislocation.

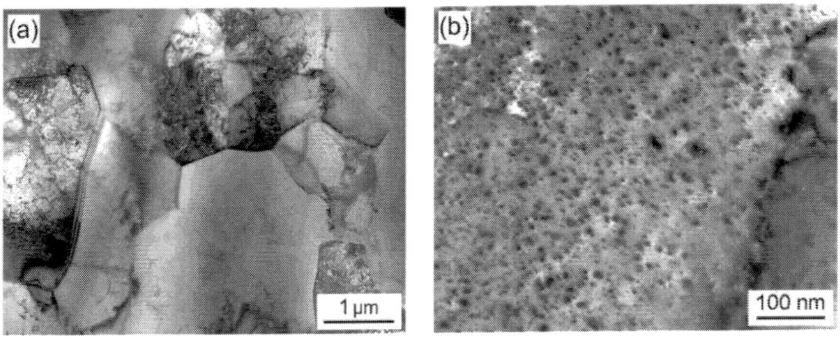

Figure 13. Microstructure of Al-0.2wt.%Sc alloy after 8 ECAP passes and subsequent creep exposure at 473 K and 20 MPa. (a) microstructure, and (b) precipitates Al_3Sc.

The EBSD data indicate that the number of high angle-boundaries (θ>15°) measured in the specimens after ECAP and subsequent creep exposure is strongly dependent on the number of ECAP passes. The number of high-angle grain boundaries is increasing with increasing number of ECAP passes from approximately 2% in the specimen after 1 ECAP pass and subsequent creep to ~ 70% in the specimen after 8 ECAP passes and subsequent creep. It was reported [4] that the grain boundary sliding can occur in UFG materials at elevated temperatures. Thus we can suppose that changes in the number of high-angle grain boundaries in the microstructure of ECAPed materials during creep tests can affect their creep behaviour by increasing the contribution of grain boundary sliding to the total creep strain [27].

The EBSD analyses were performed on the several places of the gauge length of creep specimen after ECAP and subsequent creep revealed scatter in the number of high angle grain boundaries (HAGBs). In the Figure 14 the minimal and maximal measured values of the number of HAGBs are plotted. The inspection of Figure 14 shows that the scatter in HAGBs can be particularly expected after creep tests in the specimens with lower number of ECAP passes. The heterogeneous distribution of HAGB can probably influence the homogeneity of grain boundary sliding. In the areas with the higher number of HAGBs the grain boundary sliding will be more intensive than in the surrounding areas [8].

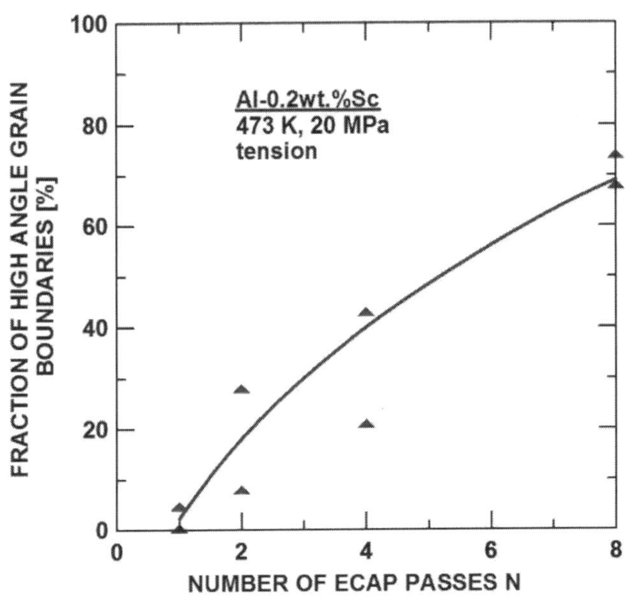

Figure 14. Fraction of high angle grain boundaries as a function of the number of ECAP passes in the Al-0.2wt.%Sc alloy.

The investigation of the unetched surfaces of the specimens after 2-8 ECAP passes and after creep exposure revealed the appearance of mesoscopic shear bands [14,15,35, 40-42] lying near to the shear plane of the last ECAP pass (Figure 15). On the surface of specimens the mesoscopic shear bands were particularly observed near the fracture region and their frequency decreased rapidly with increasing distance from the fracture. On the specimen surface after 8 ECAP passes the mesoscopic shear bands already covered almost the whole gauge length. It was found that the width of the bands decreases with increasing number of ECAP passing and after 8 ECAP passes the average width of the bands was ~ 35 μm as it is shown in Figure 15. The analyses of microstructure on the interfaces of the bands found that in the vicinity of these interfaces high heterogeneity in the distribution of HAGBs can be observed (Figure 15). The formation of the mesoscopic shear band can be related to inhomogeneity of microstructure of ECAPed alloy after creep exposure. Examination by EBSD revealed that the microstructure of mesoscopic shear bands is created by high-angle grain boundaries (Figure 15 and16).

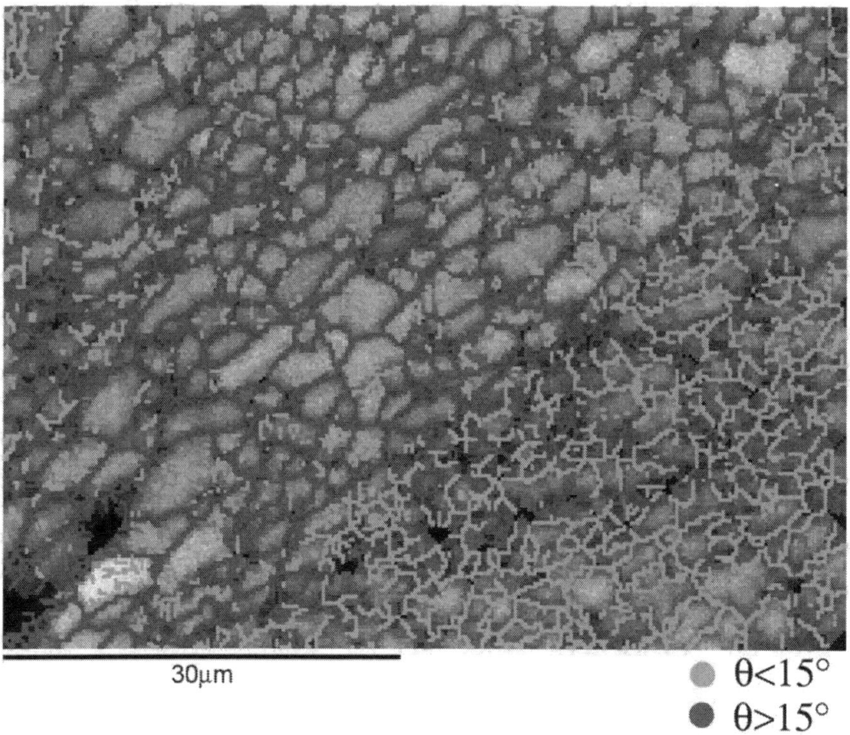

Figure 15. The heterogeneous distribution of HAGBs (red coloured) in the sample of an Al-0.2wt.%Sc alloy after 4 ECAP passes and creep at 473 K and 20 MPa.

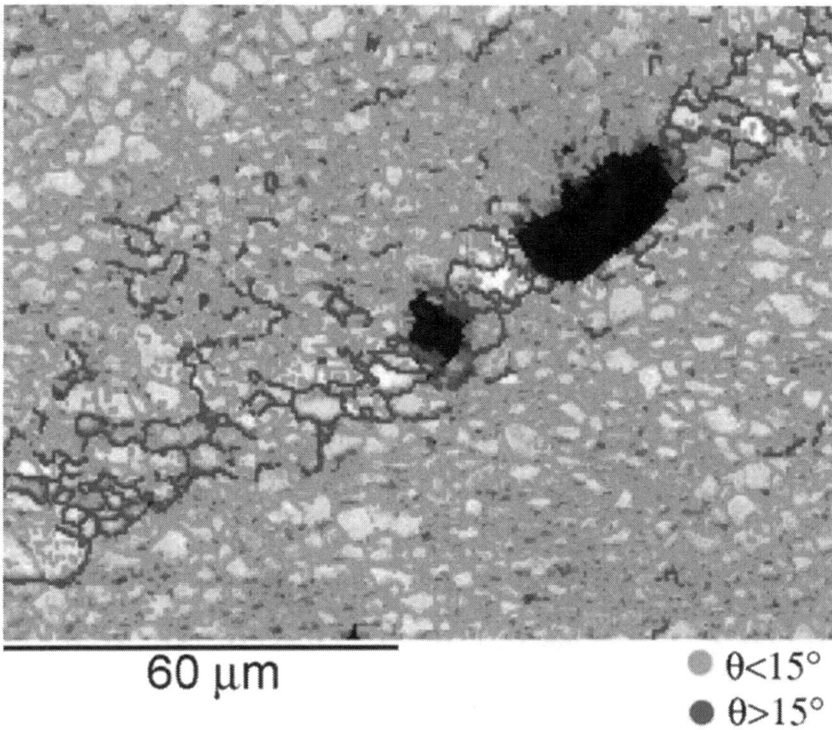

Figure 16. Appearance of the microstructure in the Al-0.2wt%Sc alloy after 2 ECAP passes and subsequent creep at 473 K. Tensile axis is horizontal, SEM.

3.3. Unique Features Of Microstructure In Ultrafine-Grained Materials

The processing technique used to obtain the UFG microstructure should strongly influence the creep properties of the material. This is primarily due to difference in microstructure as distribution of grains, subgrains, dislocation density and boundary character. The grain boundary character is usually quantified using the misorientation angle θ across grain boundaries, with high and low angle grain boundaries defined as $\theta \geq 15°$ an $2° < \theta < 15°$, respectively. Electron back scatter diffractions (EBSD) mapping has been used to quantitatively characterize boundaries in UFG materials [10,20,21,23,24]. Although the boundary spacing saturates after the first few ECAP passes, the fraction of high angle boundaries continues to increase with increasing ECAP passes [27,34,43]. In addition to grain size determination, there are a number of important microstructural parameters evaluated from EBSD but not available from conventional methods of grain characterization in particulars parameters relating to the grain orientations and boundary characters [44 - 46]. The following text describes representative results of quantitative characterization of UFG microstructure.

3.3.1. Stereological Estimates Of Ufg Microstructure Characteristics

At each examined specimen there were made three mutually perpendicular planar metallographic sections denoted as XY, XZ (longitudinal sections) and YZ (transverse section), where X, Y and Z are the axes of the Cartesian coordinate system with X along the last pressing direction and Z perpendicular to the bottom of the channel. The technique of automated EBSD in the scanning electron microscope was used for quantitative metallography. Four ranges of the boundary misorientation Δ were selected; $2° \leq \Delta$, $5° \leq \Delta$, $10° \leq \Delta$ and $15° \leq \Delta$. Then standard intercept counting [45] resulting in the mean number N_L of profile chords per unit length of the examined test lines was carried out. In each specimen, six systematically selected directions of the test lines in each section were examined. The mean boundary areas unit volume were then estimated by the stereological relation $S_V = [2N_L]$. Another important feature of the grain boundary structure is its inhomogeneity. The dispersion of grain profile areas can be qualified by the coefficient of variation CV_a of the grain profile areas in a plane $CV_a = \frac{\sqrt{V}}{\bar{x}}$, where V is grain profile areas variation and $\bar{x}$ is the mean value of the grain profile area [22,47,48]. The coefficient of variation CV_a of the profile areas is perhaps the best stereometric characteristic to evaluate homogeneity of microstructure and nowadays it is relatively easily attainable by a computer image analysis [49].

3.3.2. Inhomogeneity Of Ufg Microstructure

There are numerous reports of the processing of various pure metals and metallic alloys by ECAP and many of these reports involve a detailed characterization of the microstructure. These results are summarized in recent reviews [3,4]. However, information seldom is reported on the percentage of high angle grain boundaries (HAGB´s), an important parameter in the comparison of plasticity of different processing routes and materials [50]. It can be expected that samples with different distributions of misorientation across the grain boundaries will deform differently. Further, to provide information on the optimum microstructure of UFG materials we need to use an additional quantitative microstructural parameter other than just the average grain size critical for the creep behaviour and properties

[51]. Such parameter could be a coefficient of profile CV_a as a measure of homogeneity of materials microstructure [48].

Hence, the grain and subgrain structure of the creep specimens was revealed by means of EBSD and characterized by the coefficient of variation CV_a of the profile areas. Four ranges of the boundary misorientation Δ between adjacent pixels were selected for examination using EBSD, which correspond namely to subboundaries, transitive and high angle grain boundaries within $2° \leq \Delta$ and $5° \leq \Delta$, transitive and high angle grain boundaries for $10° \leq \Delta$, and mostly grains with HAGB´s for $\Delta \geq 15°$. Selected examples of images of XZ sections produced by EBSD of an Al-0.2wt.%Sc are shown in Figure 17.

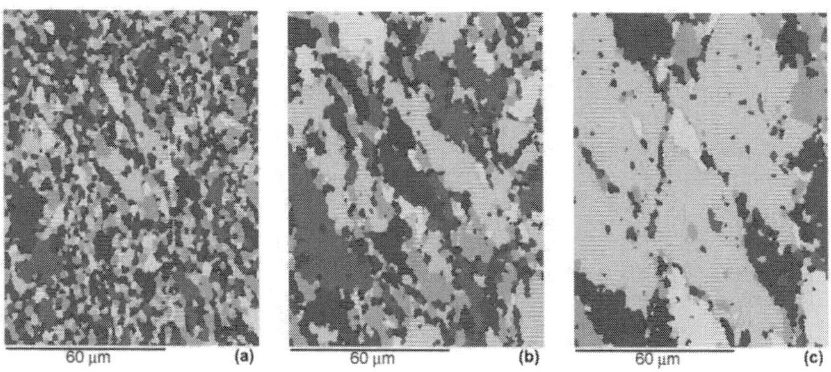

Figure 17. Selected examples of EBSD grain maps of an Al-0.2wt.%Sc alloys after ECAP constructed for different grain misorientations: (a) subboundaries $\Delta \geq 2°$, (b) transitive subboundaries and high angle boundaries $\Delta \geq 10°$, and, (c) high-angle grain boundaries $\Delta \geq 15°$.

It was generally observed that with the increasing number of ECAP passes N, a considerable amount of subgrain boundaries was gradually transformed to HAGB´s as shown in Figure 18. At the same time, the local homogeneity of structure as characterized by the values of CV_a with the increasing number N is gradually improved as demonstrates by Figure 19. Figure 19 also shows the microscopic appearance of the specimens of an Al-0.2wt.%Sc alloy crept under the same loading conditions (473 K, 20 MPa) but processed by different numbers of ECAP passes N. The values of CV_a as high as 10 at N = 2, $1 \leq CV_a < 2$ at N = 4, and $0.55 \leq CV_a \leq 1$ at N = 8 were found. Extremely high value CV_a at N = 2 demonstrates very high inhomogeneity of a mixture of subgrain and grain structures. The value of CV_ain very homogeneous grain systems should not exceed the value of 1 [23]. It should be noted that extremely high values of the coefficient of variation CV_a is a natural consequence of the short as well as long-range inhomogeneity of microstructure.

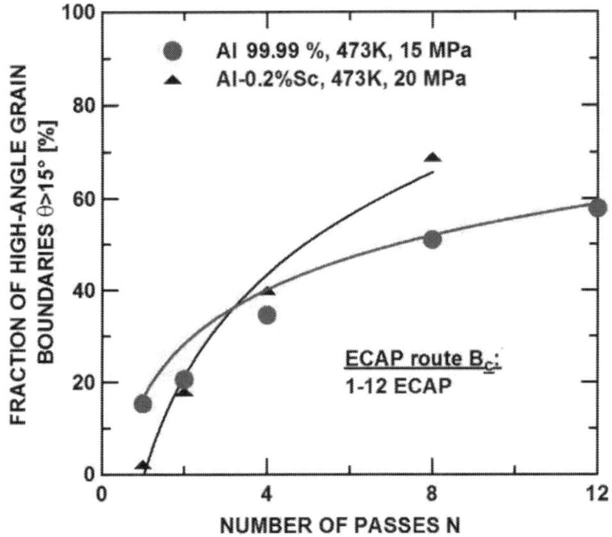

Figure 18. The fraction of high-angle grain boundaries in the crept samples as a function of the number of ECAP passes.

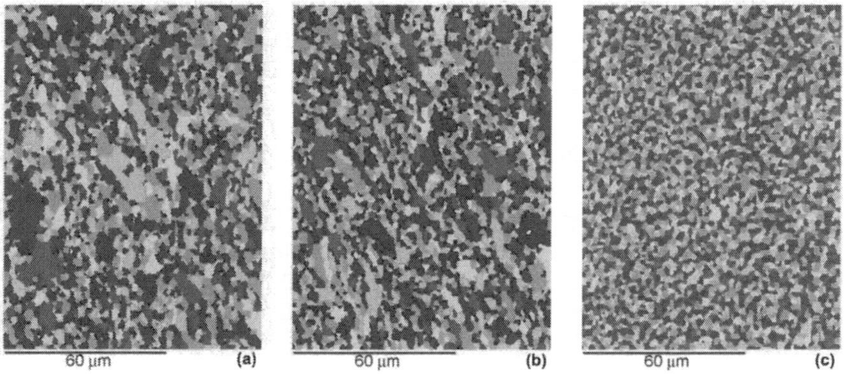

Figure 19. Grain maps of an Al-0.2wt.%Sc alloy processed by different number N of ECAP passes and crept at 473 K and 20 MPa, and corresponding parameter CV_a: (a) 2 passes, $CV_a >> 2$, (b) 4 passes, $CV_a < 2$ and, (c) 8 passes, $0.55 \leq CV_a < 1$.

The substantial grain coarsening especially in the case of pure metals came up during the creep exposures depending on stress and temperature thus manifesting the thermal instability of ultrafine-grained microstructure [4,52,53]. It is clear that in pure aluminium the grains grow rapidly at elevated temperatures because there are no precipitates within the crystalline lattice to restrict the movement of the grain boundaries by a "pinning effect". By contrast, submicrometer grains may be retained to relatively high temperatures in materials containing a distribution of fine precipitates as in the case of an Al-0.2%Sc alloys containing Al_3Sc precipitates [29]. The quantitative characterization of the inhomogeneity of the boundary structure is shown in Figure 20. It should be stressed that the values of the coefficient of profile area CV_a as a measure of structure homogeneity strongly depend on the chosen ranges of misorientation Δ in EBSD analysis. Whereas the fraction of the subboundaries (low-angle grain boundaries) are dominating for $\Delta \geq 2°$ (Figure 21a), the fractions of high-angle grain boundaries ($\theta \geq 15°$) confirm their high share for the range of $\Delta \geq 15°$ (Figure 21b). Detailed inspection of Figure 20b shows a strong dependence of CV_a after creep on the number of ECAP passes. Substantial decrease of the values CV_a 2° for the precipitation-strengthened Al-0.2Sc alloy with increasing number of passes N for $\Delta \geq 2°$ may be connected with the more rapid evolution boundaries having misorientation angles $\theta > 15°$ (Figure 20a).

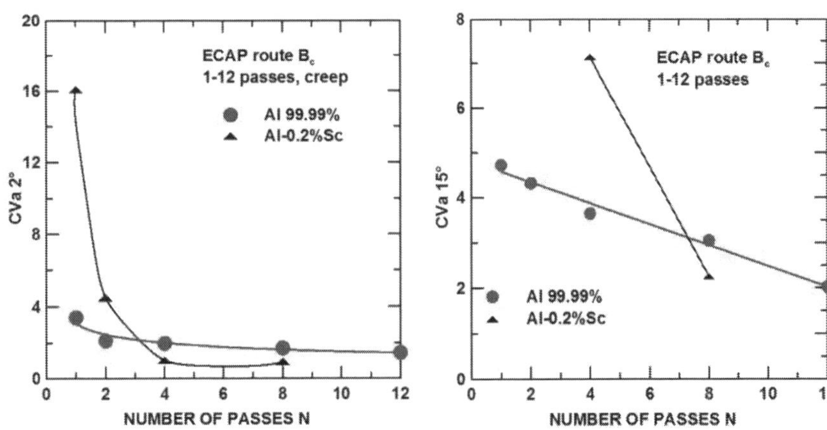

Figure 20. Coefficient of profile area CVa as a measure of homogeneity: $0.55 \leq CVa < 1$ (homogeneous system), and $CVa >> 2$ (multimodal grain size distribution). The chosen ranges of misorientation Δ in EBSD analysis: (a) $\Delta \geq 2°$, (b) $\Delta \geq 15°$.

4. CREEP BEHAVIOUR OF UFG ALUMINIUM AND ITS ALLOYS

The mechanical properties of bulk ultrafine-grained (UFG) materials at elevated and/or high temperatures are a new and important area of research [4]. However, there have been only a few investigations on the creep behaviour of bulk UFG materials processed by equal-channel angular pressing (ECAP) [43,54]. By comparison with the unpressed (coarse-grained) state, processing by ECAP may lead to considerable changes in the creep properties in bulk UFG materials including a decrease and/or an increase [8] in the minimum creep rate and the creep life. It is important to note that these trends may be noticeably dependent on the number of ECAP passes. Many investigations concerned with the identification of creep mechanisms have been undertaking using coarse-grained pure aluminium [55] and commercial aluminium alloys [56]. However, it is logical to expect that the mechanism of hardening/softening observed in the aluminium processed by ECAP may be different from that observed in the coarse-grained material. Consequently, it cannot be excluded that creep in the ECAP aluminium and its alloys is controlled by different creep mechanism(s) than that in the coarse material. Thus, neither phenomenological nor microscopic aspects of the creep behaviour of materials processed by ECAP have been understand sufficiently as yet.

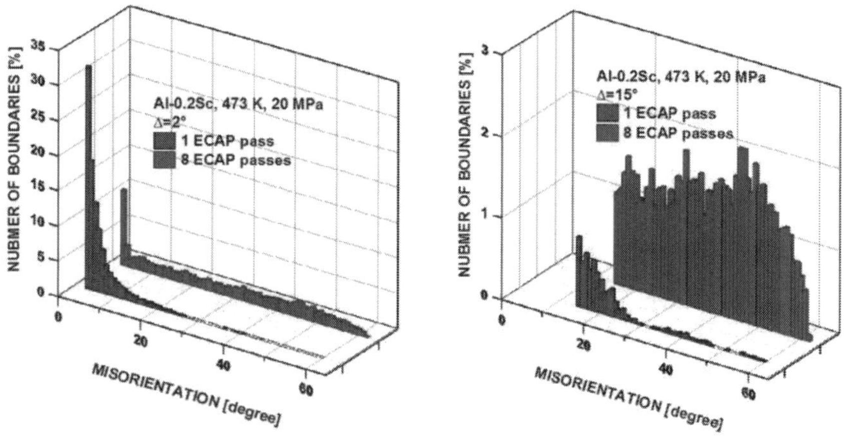

Figure 21. Distribution of boundaries with different misorientation θ for an Al-0.2wt.%Sc alloy analysed inFigure 8: (a) EBSD analysis for Δ ≥ 2°, (b) Δ ≥ 15°.

This section reports a series of creep experiments that were conducted on specimens of pure aluminium and its Al-0.2wt.%Sc and Al-2wt.%Mg-0.2wt.%Sc alloys processed by equal-channel angular pressing. For comparison purposes, some creep tests were performed also on the unpressed materials. Creep tests were performed both in tension and compression.

4.1. Effect Of Processing Route On Creep Behaviour

As discussed in more detail in Section 2.2. four distinct pressing routes have been identified (Figure 3). The ECAP processing was conducted by one or repetitive passes following either route A, B (route B_C was used only) or C. Figure 22 contains a complete record of the main creep parameters for the ECAP specimens of Al after creep testing in tension at 473 K and 15 MPa (each point represents the average results of two to three individual creep tests at the same loading conditions). Inspection ofFigure 22 shows there are not very significant differences in creep properties of specimens prepared by the various ECAP processing routes. All three processing routes produce a significant increase in the minimum creep rate through the first four passes and a slight increase during subsequent pressing (Figure 22a). By contrast, the time to fracture (creep life) dramatically drops through four passes and then there is no significant differences among the number of following passes – Figure 22b.

Recently, attention has been given to effectiveness of the various ECAP routes in producing grain refinement in aluminium [10]. It has been demonstrated that ECAP is capable of producing refined structures with large fractions of high-angle boundaries [10] although the mechanisms involved in the formation of fine grains and high-angle boundaries in the deformation microstructure remain to be clarified. In this work microstructural investigation all routes examined indicated little differences in the grain size produced via the various ECAP routes. With increasing number of ECAP passes this difference decreases. Further, there was little apparent dependence of the misorientation on the various process route for an ECAP die having an internal angle equals to 90°. The misorientation data confirmed that repetitive pressing results in a progressive increase in the fraction of high-angle grain boundaries (Figure 18).

In related work, Sklenicka et al. [11,57] carried out an extensive creep testing on pure aluminium processed by various ECAP routes. It was found that processing route had a little apparent effect on the creep behaviour of a pressed aluminium. However, the effect of grain growth during creep may tend to obscure the effect of different processing routes and the creep experiments are probably not a sufficiently refined procedure for picking up these rather small differences in the creep behaviour.

4.2. Creep Behaviour Of Pure Aluminium

Creep tests were performed on the as-pressed specimens both in tension and compression in the temperature interval from 423 to 523 K under an applied stress range between 10 and 25 MPa. The subsequent ECAP passes were performed by route B_C (see part 2.2.) up to 12 passes.

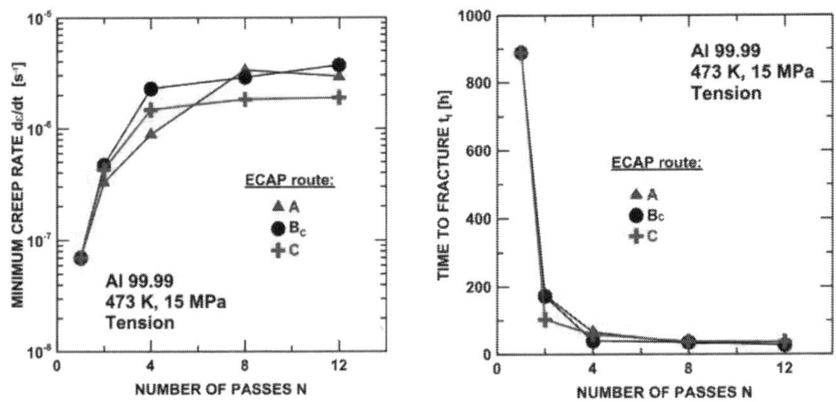

Figure 22. Influence of different ECAP routes and different number of ECAP passes on (a) creep rate, and (b) time to fracture.

4.2.1. Creep Behaviour

Representative creep curves are shown in Figures 23 and 24. All of these plots were obtained at temperature of 473 K (~ 0.5 T_m) under an initial applied uniaxial tensile or compression stress of 15 MPa. The creep tests in tension were run up to the final fracture of the creep specimens, whereas the creep tests in compression were interrupted at a true strain of about ~ 0.35.

Standard ε vs t creep curves in Figures 23a and 24a can be easily replotted in the form of the instantaneous strain rate dε/dt versus strain as shown in Figures 23b and 24b. As demonstrated by figures, significant differences were found in the creep behaviour of the ECAP material when compared to its coarse-grained counterpart. First, the ECAP materials exhibits markedly longer creep life (Figure 23a) or markedly longer duration of creep exposure to obtain a strain of ~ 0.35 (Figure 24b) than coarse grained aluminium. Second, the minimum creep rate for the ECAP material is about one to two orders of magnitude less than that of coarse-grained material. Third, the shapes of tensile creep curves for the ECAP material after high number of pressing differ considerably from the tests conducted at small number of the ECAP passes by the extent of individual stages of creep.

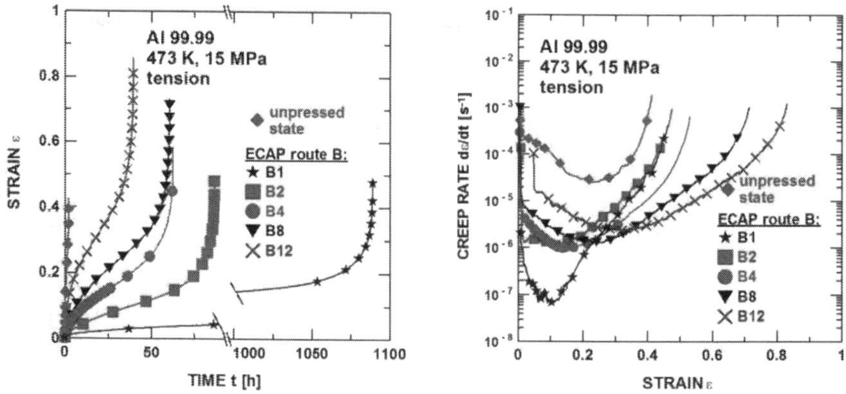

Figure 23. Standard creep and creep rate versus strain curves for unpressed state and various number of ECAP passes via route B_c (creep in tension up to fracture).

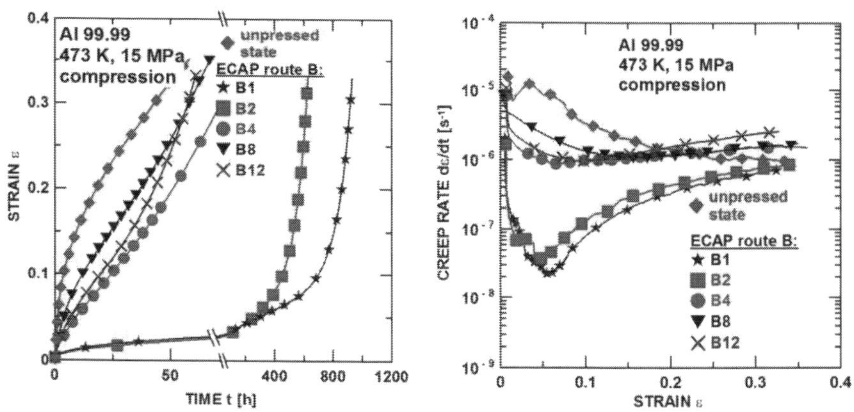

Figure 24. Standard creep and creep rate versus strain curves for unpressed state and various number ECAP passes via route B_c (creep in compression up to strain ~ 0.35).

The difference in the minimum creep rate for the ECAP material and unpressed state consistently decreases with increasing number of ECAP passes (Figures 23b and 24b). An additional difference is illustrated by Figure 25a, which shows the variation of the minimum creep rate with the applied stress for the ECAP specimens after 8 passes. The results demonstrate that at high stresses the minimum compressive creep rate of the ECAP material may be up to one order of magnitude lower than that of the unpressed material, although this difference decreases with decreasing applied stress and becomes negligible at 10 MPa. The observed values of the stress exponent $n = (\partial \ln \dot{\varepsilon} / \partial \ln \sigma)_T$ are ~ 6.6 for the unpressed material, ~ 4.8 (creep in compression) and ~ 5.7 (creep in tension) for the ECAP Al, respectively.

To determine the apparent activation energy for creep Q_c, the minimum creep rate was measured in the temperature interval from 423 to 523 K and at two tensile applied stresses 15 and 20 MPa, respectively. The activation energy for creep Q_c is defined as

$$Q_c = \left[\frac{\partial \ln \dot{\varepsilon}_{min}}{\partial (-1/kT)} \right]_\sigma \qquad (1)$$

Thus, the activation energy Q_C can be derived from the slope of log dε/dt versus 1/T plots shown inFigure 25b. A value of the apparent activation energy, Q_C, was determined by the least square methods. The Q_C is stress dependent and equals to 129.7 ± 16 and 110.9 ± 9 kJ/mol for stresses 20 and 15 MPa, respectively.

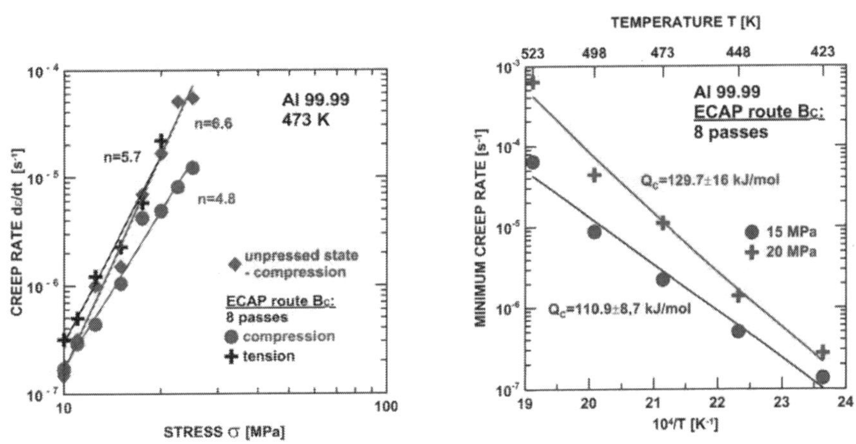

Figure 25. Dependence of minimum creep rate for unpressed state and 8 ECAP passes on: (a) applied stress, (b) testing temperature at two levels of stress.

4.2.2. Grain Boundary Sliding (Gbs)

The total creep strain ε generally consists of following contributions: the strains caused by dislocation glide and nonconservative motion of dislocations, grain boundary sliding (GBS) [58-60], stress directed diffusion of vacancies and by intergranular void nucleation and growth, respectively. However, it should be noted that not all of the above processes operating are independent of each other, as frequently assumed. A possible explanation for the occurrence of intensive GBS in UFG materials is that diffusion is more rapid in ECAP processed materials with highly non-equilibrium grain boundaries [4, 19, 61,62]. Accordingly, it appears that GBS is easier in these UFG materials.

The amount of grain boundary sliding (GBS) was determined by measuring the surface offsets produced at the intersections of grain boundaries with marker lines transverse to the stress axis [27,58].Figure 26 shows one clear example of the occurrence of grain boundary sliding in creep of the ECAP aluminium. Longitudinal displacements of the marker lines, u, due to GBS, together with the fraction of boundaries, κ_S, with observable GBS, were measured using SEM.

Grain boundary sliding was measured on the surfaces of the tensile specimens crept up to a predetermined strain $\varepsilon \approx 0.15$. Scanning electron microscopy made it possible to detect GBS characterized by $u \geq 0.1\mu m$. However, GBS was not observed at all grain boundaries; that is why the relative frequency of sliding boundaries κ_s was determined. Then the strain component ε_{gb} due to GBS is expressed as [27,58]:

$$\varepsilon_{gb} = (1 + \varepsilon)\,\bar{u}.\kappa_s / \overline{L} \qquad (2)$$

where the mean grain size $\overline{L}$ was determined by the linear intercept method and the overall contribution of GBS to the total creep strain in the specimen, γ, was estimated as $\gamma = \varepsilon_{gb}/\varepsilon$. The results of GBS measurements are summarized in Table 3. It is evident that the fraction of boundaries $\kappa_s$increases as the number of ECAP passes increases. This result supports the idea that GBS is connected with microstructural changes of grain boundaries [27]. It is to note that in the best case (12 passes) the contribution of GBS to creep strain is only 33%.

4.2.3. Creep Deformation Mechanisms

The mechanisms controlling the creep properties of pure metals have been usually identified from the dependence of the minimum and/or steady-state creep rate $\dot{\varepsilon}_m$ on stress σ, absolute temperature T and grain size d, using a power-law expression of the form

$$\dot{\varepsilon}_m = A\sigma^n (1/d)^p \exp(-Q_c/RT) \quad (3)$$

Table 3. Summary of GBS measurements (ε ≅ 0.15).

No of Passes	ū[μm]	κ_s	$\bar{L}$ [μm]	ε_{gb}.10[2]	$\varepsilon_{gb}/\varepsilon$.10[2]
1	0.51	0.80	14.9	3.15	21.0
2	0.48	0.83	12.7	3.60	24.0
4	0.55	0.93	12.2	4.80	32.0
8	0.49	0.91	10.8	4.70	31.0
12	0.52	0.92	11.0	5.00	33.0

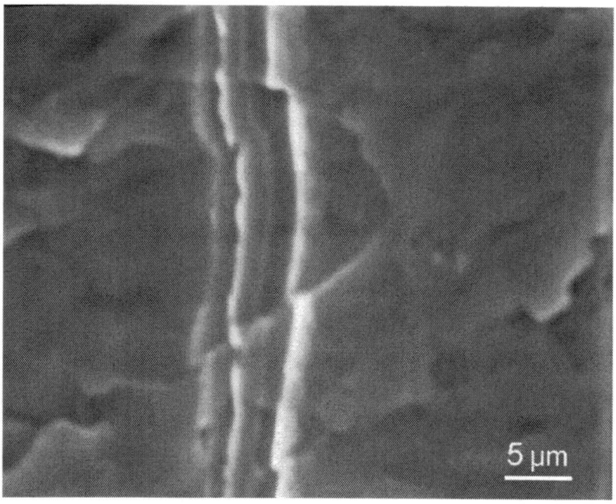

Figure 26. Example of grain boundary sliding in the ECAPed aluminium (route B_c, 8 passes) after creep testing at 473 K and 15 MPa. Tensile stress axis is horizontal.

where Q_c is the activation energy for creep. With this approach, the fact that n, p and Q_c are themselves functions of stress, temperature and grain size is conventionally explained by assuming that different mechanisms, each associated with different values of n, p and Q_c, control the creep characteristics in different stress/temperature regimes. In turn, the dominant mechanisms under specific test conditions are then generally determined by comparing experimentally determined values of n, p and Q_c with the values predicted theoretically for different creep mechanisms.

In Figure 27, the minimum creep rates $\dot{\varepsilon}_{min}$, are plotted against applied stress σ for pure aluminium using the data presented in Figure 25a. Experimental points are shown for both the unprocessed (coarse-grained) and for the UFG aluminium after 8 ECAP passes (only the results of tensile creep testing are used).

The broken lines in Figure 27 denote the model predictions of the theoretical creep rates according models for various creep deformation mechanisms, namely for superplastic flow, Nabarro-Herring [59,63,64] and Coble [59,65] diffusion creep and power-law creep by dislocation climb and glide processes. It should be stressed that the phenomenon of dislocation diffusion is not well understood on the fundamental level at present. The theoretical creep rates were calculated from equation (3) using the material data presented in Table 4.

Table 4. Creep mechanisms and material data.

Al (473K) d_{ECAP} = 1μm d_{CREEP} = 12μm [27]								
number of curve	**creep mechanism**	**n**	**p**	**A**	**D [$m^2 s^{-1}$]**		**Q [$kJ mol^{-1}$]**	**source**
(1), (1*)	superplastic flow	2	2	10	D_{GB}	5.9×10^{-14}	86	[66]
(2), (2*)	Nabarro – Herring creep	1	2	28	D_L	2.72×10^{-20}	143.4	[66]
(3), (3*)	Coble creep	1	3	62	D_{GB}	5.9×10^{-14}	86	[66] [67]
(4)	dislocation climb and glide	5	0	10^3	D^*	1.9×10^{-14}	124	[66]

D* is the effective diffusion coefficient which incorporates contribution from both lattice and grain boundary diffusion, D_{gb} and D_L are the grain boundary and lattice diffusion coefficients, respectively.

D^* is the effective diffusion coefficient which incorporates contribution from both lattice and grain boundary diffusion, D_{gb} and D_L are the grain boundary and lattice diffusion coefficients, respectively.

It should be noted that grain growth occurs easily at the elevated temperatures used in creep experiments of pure metals. Indeed, the occurrence of significant grain growth in creep tests conducted on high-purity aluminium processed by ECAP at room temperature was observed [27,36]. Accordingly, two sets of the predicted theoretical rates were calculated in this analysis for both states of materials using the measured grain size after ECAP processing (d_{ECAP}) and after subsequent creep exposure (d_{CREEP}) [27]. The theoretical rates using d_{ECAP} will be marked by simple number (e.g. 1) while a numbering with asterisk (e.g. 1*) will be used for the rates corresponding to d_{CREEP} in Figure 27.

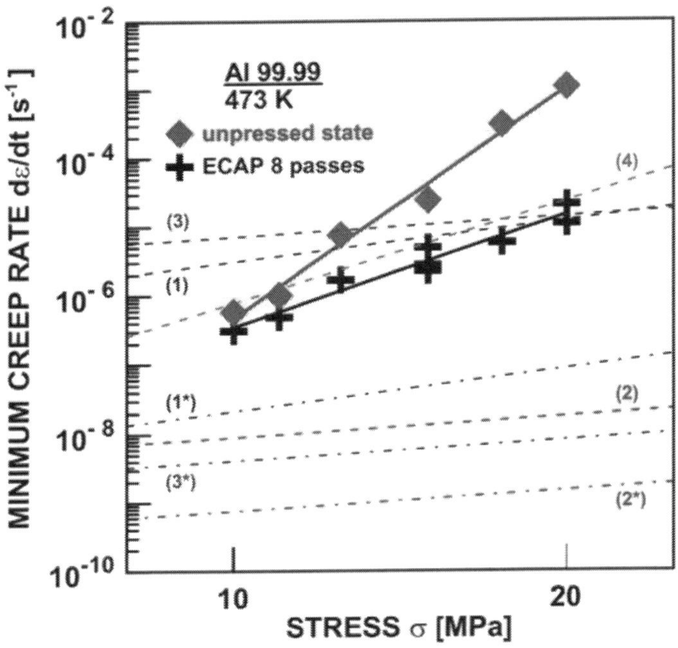

Figure 27. Experimentally determined and theoretically predicted the stress dependences of the minimum creep rates for various creep mechanisms in aluminium.

Figure 27 demonstrates that at high applied stresses the experimentally determined minimum creep rate of the ECAP aluminium may be up to two orders of magnitude lower than that of the unpressed material, although this difference decreases with decreasing applied stress and becomes nearly negligible at 10 MPa. The predictions show that under the creep loading conditions investigated Nabarro-Herring and Coble diffusion creep and superplastic flow are too slow to account for the creep deformation considering a significant grain growth in the pressed materials. Also shown in Figure 27are the predicted theoretical creep rates for ultrafine-grained states (d_{ECAP}) after ECAP which are within two to five orders of magnitude faster than that for the creep of coarsened materials. However, such predictions are not correct a priori due to thermal instability of microstructure of the pressed materials. Inspection of Figure 27 shows that for the pressed material there is an excellent agreement between the experimental datum points and the predicted creep behaviour based on dislocation climb and glide. Further, for $n \geq 4$ creep is known to occur by diffusion-controlled movement of dislocations within grains and/or along grain boundaries (grain boundary sliding). The high value of n_{CG} for the unpressed aluminium (Figure 27) could represent a regime leading into a power-law breakdown (PLB) region at rapid strain rates and/or high stress levels.

The analysis of Figure 27 indicates that creep in pure aluminium after ECAP occurs by the same mechanism as in conventional coarse-grained materials with intragranular dislocation glide and climb as the dominant rate-controlling flow process. Therefore, the activation energy for creep Q_C should be the same as the value of the activation enthalpy of lattice self-diffusion ΔH_{SD} ($\Delta H_{SD} \cong 127 - 143$ kJ/mol in aluminium [59]). The values obtained for Q_C (Figure 25b) are somewhat lower than that of ΔH_{SD}. Supposing that grain boundary sliding is controlled by grain boundary diffusion, which is assumed to be about 0.7 times that for lattice self-diffusion, the presented results give support to the assumption that GBS may be increasingly important in creep of the ECAP aluminium at low applied stresses.

It can be concluded that the creep resistance of high purity aluminium is increased considerably already after one ECAP pass. However, successive ECAP pressing leads to a noticeable decrease in the creep resistance. The results of microstructure investigations indicate that an inhomogeneity and thermal instability of the ECAP microstructure may strongly influence the creep behaviour of the pressed material [48].

4.3. Creep Behaviour Of Aluminium Alloys

The combination of solid-solution strengthening and precipitation strengthening in creep of aluminium alloys at elevated temperatures has not been extensively studied. Numerous reports dealt with the creep behaviour of Al-Mg solid solution [68-70]. Most precipitation strengthened aluminium alloys currently being used are limited to relatively low temperature usage, because of the dissolution and/or rapid coarsening of their precipitates. An exception represents Al-Sc alloys containing low volume fractions of very fine coherent cuboidal Al_3Sc precipitates. The dislocation creep behaviour of coarse-grained binary Al-Sc alloys at 573 K and the precipitation strengthening effect of the Al_3Sc phase were investigated by Fuller et al. [71] and Seidman et al. [72]. Recently, the effect of Mg addition on the creep behaviour of an Al-Sc alloy was reported by Marquis et al. [31]. It was found that the creep strength of an Al-3wt.% Mg-0.2wt.%Sc alloy, containing Mg in solid solution and Al_3Sc as nanosize precipitates, is significantly improved compared to binary Al-Sc alloys.

4.3.1. Creep Behaviour

As it was reported in Section 4.2. the processing by ECAP of a coarse-grained high purity aluminium provided a potential for marked improvement in the creep properties. Accordingly, Section 4.3. reports on a systematic study of the creep behaviour of the ECAP processed aluminium alloys containing low volume fraction of Al_3Sc precipitates to elucidating the effect of ECAP on their creep resistance.

Figure 28a shows standard strain ε versus time t curves for the as-received (unpressed) Al-0.2wt.%Sc and Al-3wt.%Mg-0.2wt.%Sc alloys and those for the same alloys processed by ECAP through 8 passes at 473 K and 50 MPa (an exception is 80 MPa for Al-3wt.%Mg-0.2wt.%Sc alloy in the as-received state). The creep tests in compression were interrupted at a true strain of about ~ 0.35. These standard creep curves were replotted in the form of the instantaneous creep rate dε/dt versus time t as shown inFigure 28b. It is clear that no-one of the creep curves exhibits a well-defined steady state. In fact this stage is reduced to an inflection point of the dε/dt versus t curve. Supposing that the instantaneous creep rate dε/dt at given stress and temperature is a certain measure of the "softness" of the microstructure, then the dε/dt-t plots reveal the time evolution of this "softness". However, the dε/dt-ε plots may give additional information, since they reflect the effect of the plastic creep strain on the instantaneous "softness" of the microstructure (Figure 28c).

The differences in the minimum creep rates for pure Al, Al-0.2wt.%Sc and Al-3wt.%Mg-0.2wt.%Sc in the as-received and as-pressed conditions are illustrated most readily in Figure 29 showing the variation of the minimum creep rate with applied stress. The results demonstrate that for pure aluminium at high stresses the minimum creep rate of ECAP material may be up to one order of magnitude lower than that of the unpressed material, although this difference decreases with decreasing applied stress so that, at 10 MPa is negligible. By contrast, when tests of Al-Sc and Al-Mg-Sc alloys are performed at the same stress, the creep rates in the as-pressed alloys are faster than in the unpressed alloys by more than two and/or three orders of magnitude on the strain rate scale. The stress dependence of the minimum creep rate for the as-pressed Al-Mg-Sc alloy at lower stresses ($\sigma < 20$ MPa) is different in trend, which is clearly demonstrated by the characteristic curvature on the plot inFigure 29.

Such behaviour is generally associated with the presence of a threshold stress marking a lower limit stress below which no measurable strain rate can be achieved [73,74]. The threshold stress value determined from linear plot [75] of $\dot{\varepsilon}^{1/n}$ versus σ is about 6 MPa for the alloy studied. The validity of this approach is illustrated in Figure 30, where the data fall on straight line with the slope of 3, consistent with the assumed value of ~ 3 (Figure 29) for the stress exponent of the minimum creep rate $n = (\partial \ln \dot{\varepsilon}_{min} / \partial \ln \sigma)_T$.

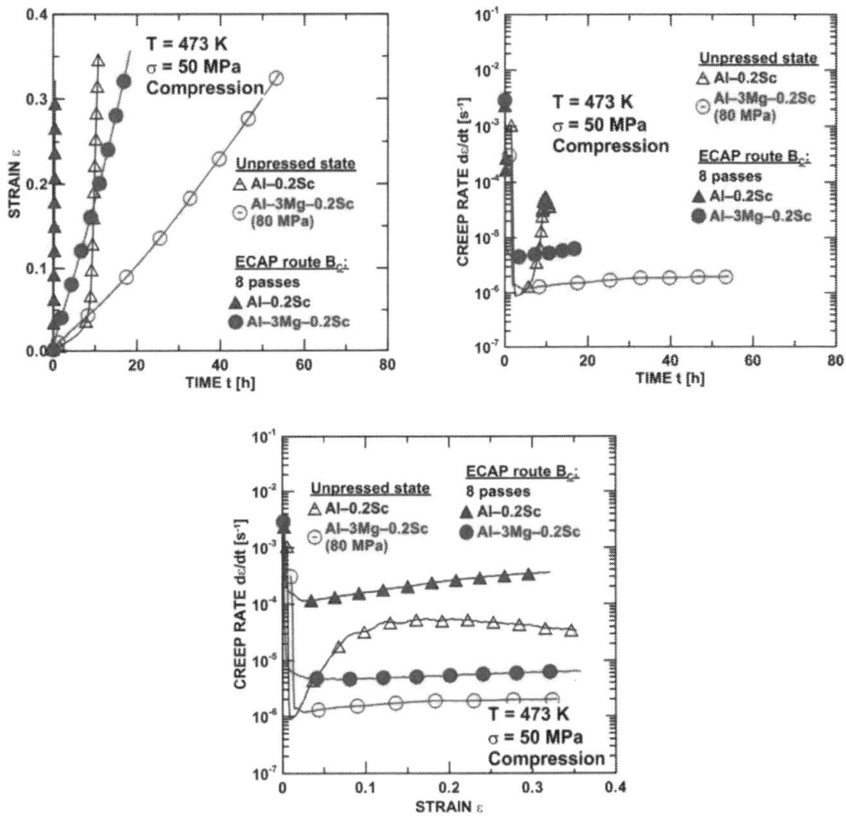

Figure 28. Creep curves for specimens after ECAP processing through 8 passes and for unpressed specimens: (a) standard creep curve, (b) creep rate vs. time, (c) creep rate vs. strain.

Recently the tensile creep experiments on the same binary Al-0.2%Sc alloy were reported [38]. The stress dependences of the minimum creep rates and the times to fracture for this alloy after 1 and 8 ECAP passes are shown in Figure 31. As demonstrates by Figure 31, the pressed alloy after 8 ECAP passes exhibits the very similar value of n as the results obtained by compression tests (Figure 29).

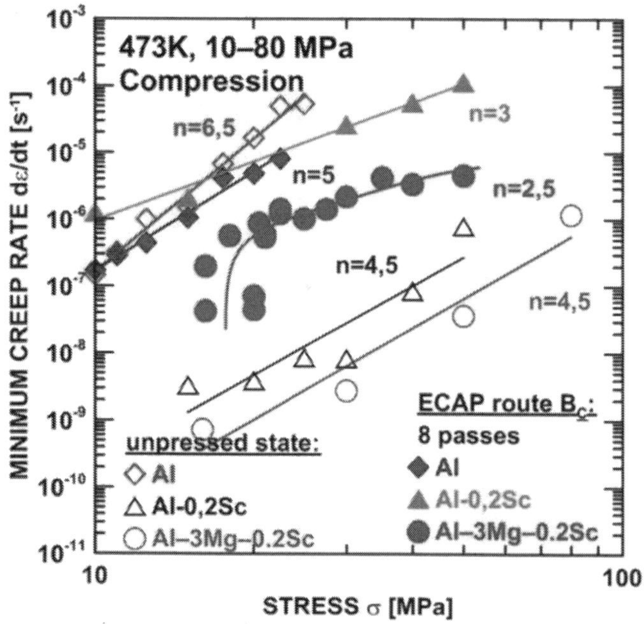

Figure 29. Stress dependences of minimum creep rate for pure aluminium and its alloys in the unpressed and ECAPed conditions.

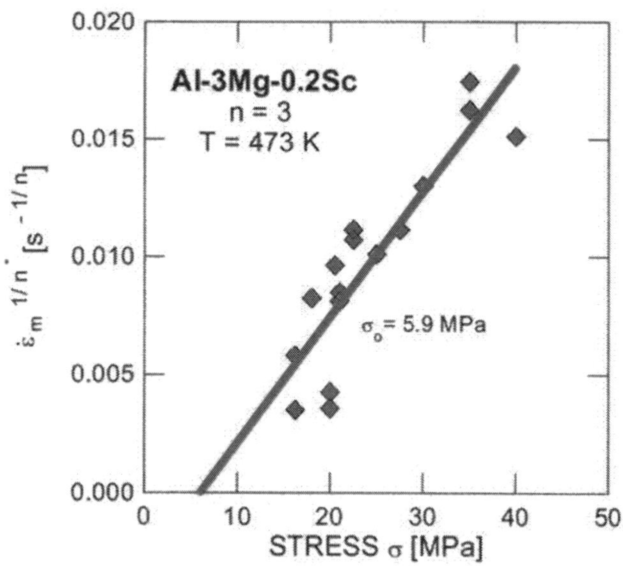

Figure 30. The linear extrapolation procedure for determining the threshold stress.

4.3.2. Creep Deformation Mechanisms

The results from this investigation on precipitation-strengthened aluminium alloys do not confirm a general validity of the conclusion of our earlier results that processing by ECAP of a coarse-grained aluminium gave a potential for an improvement in the creep resistance [26,27]. By contrast, the Al-0.2wt.%Sc and Al-3wt.%Mg-0.2wt.%Sc alloys exhibited faster creep rate than their coarse-grained counterparts when creep tested under the same loading conditions.

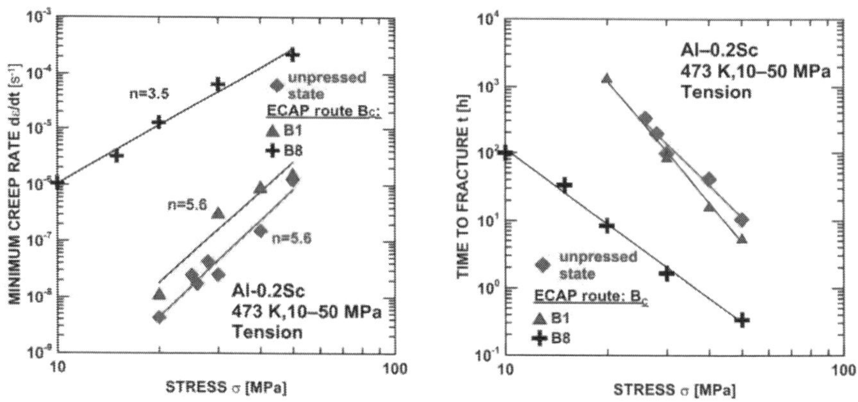

Figure 31. Stress dependence of (a) the minimum creep rates, and (b) times to fracture for Al-0.2Sc alloy.

The observed values of the stress exponents $n = (\partial \ln\dot{\varepsilon} / \partial \ln\sigma)_T$ are ~ 4.5 for the unpressed alloys and ~ 3 for the ECAPed alloys, respectively (see Figure 29). The mechanism which most probably plays the dominant role in the power-law creep (n ~ 4.5) of coarse-grained Al-0.2wt.%Sc and Al-3wt.%Mg-Sc alloys is the dislocation climb-bypass mechanism in the presence of elastic interactions between dislocations and coherent precipitates. The lower value of the stress exponent (n ~ 3) found for the ECAPed alloys may reflect the synergetic effect of more intensive grain boundary sliding in the creep of ultrafine-grained materials [27]. Thus, an important contribution of grain boundary sliding to the total creep strain in the ultrafine-grained Al-0.2Sc and Al-3Mg-0.2Sc alloys may explain the observed detrimental effect of ECAP on their creep resistance. In order to examine rigorously the differences in creep behaviour of the alloys investigated more quantitative results of microstructural analysis are needed.

The detailed analysis was undertaken by Kawasaki et al. [36] to examine the flow characteristics of the ultrafined-grained Al-3Mg-0.2Sc and Al-0.2Sc alloys [33-35,43]. The theoretical predictions of the minimum creep rates have shown that Nabarro-Herring creep is too slow to account for the creep deformation in an ultrafine-grained Al-3wt.%Mg-0.2wt.%Sc alloy [33] but there is good agreement, to within an order of magnitude, with the predictions of superplastic flow except only at the lower stresses where the points deviate from linearity and there is evidence for the presence of a threshold stress (see Figure 29).This threshold stress probably arises from the presence of coherent Al_3Sc precipitates.

For an Al-0.2wt.%Sc alloy crept in compression at 473 K [33] predictions for Nabarro-Herring diffusional creep were again too slow but there was excellent agreement between the experimental datum points and the predicted behaviour in superplastic flow. The slightly higher stress exponent after ECAP may reflect an inhibition in GBS at the lowest strain rates due to the presence of intergranular Al_3Sc precipitates. Finally, the results on an Al-0.2wt.%Sc alloy tested in creep under tensile conditions at 473 K [43] again shown that the predicted behaviour for Nabarro-Herring creep is too slow but there is reasonable agreement with the model for superplastic flow.

Thus, the results from both sets of experiments exhibit a general consistency with the predicted behaviour for conventional superplasticity. The theoretical predictions provide a clear demonstration that conventional creep mechanisms, already developed for coarse-grained materials, may be used to explain the flow characteristics of materials with ultrafine grain size. Furthemore, at least for the aluminium alloys examined in this Chapter, it is not necessary to involve any new and different creep deformation mechanisms.

4.4. Creep Ductility

Creep ductility is very important for various shaping and forming technological operations at elevated and high temperatures and especially for avoiding catastrophic failure in load-bearing parts of high temperature components. Creep strength and ductility are the key creep properties of creep-resistant materials but these properties typically have opposing characteristics. Thus, these materials may be strong or ductile but they are rarely both. In this connection, recent findings of high strength and good ductility in several bulk ultrafine-grained (UFG) metals produced by severe plastic deformation (SPD) are of special interest [4].

Typically, creep ductility in tension which can be characterized by the strain to fracture ε_f, is given by

$$\varepsilon_f(\sigma, T, S) = \int_0^{t_f(\sigma,T,S)} \dot{\varepsilon}(\sigma, T, S, t)dt \quad (4)$$

where σ is the applied stress, T is the absolute temperature, S is some parameter that characterizes the structure of specimen and t_f is the time to fracture. According to the first mean-value theorem of a definite integral eq. (4) can be expressed as

$$\varepsilon_f(\sigma, T, S) = t_f(\sigma, T, S)\dot{\varepsilon}(\sigma, T, S) \quad (5)$$

Where $\dot{\varepsilon}(\sigma, T, S)$ is some strain rate in the interval from 0 to t_f.

A strong influence of the microstructure on creep behaviour has been observed in various UFG materials [22,23,38,49,54,76]. By contrast, no report is available describing the link between microstructure and creep ductility in UFG materials processed by ECAP at elevated and high temperatures.

To investigate a course of creep strain during creep exposure and for mutual comparison of the fracture strains Figures 32a and 33a can be replotted in the form of the instantaneous strain rate dε/dt versus creep exposure time t normalized to the time to fracture t_f as shown in Figures 32b and 33b. The figures demonstrate that very significant strain contributions to the fracture strains are generated during the last tenth of creep life, however, an accumulation of the creep strains during the course of creep exposure slightly differ.

A question naturally arises about an approach to the problem of creep ductility enhancement. In this connection different approaches may be considered [4]. It has been suggested that some ductility enhancement may be associated with an increase in the fraction of high-angle grain boundaries with increasing number ECAP passes and with a consequent change in the controlling deformation mechanisms due to the increasing tendency for the occurrence of grain boundary sliding. A possible explanation for the occurrence of intensive grain boundary sliding in UFG materials is that diffusion is more rapid in ECAP processed materials with highly non-equilibrium grain boundaries [77,78]. Accordingly, it appears that grain boundary sliding is easier in these UFG materials during creeping at ambient temperatures leading to the possibility of increased ductility. Figure 34a shows the measured strain to fracture as a function of the fraction of

HAGBs when testing under creep loading conditions presented in Figures 32 and 33.

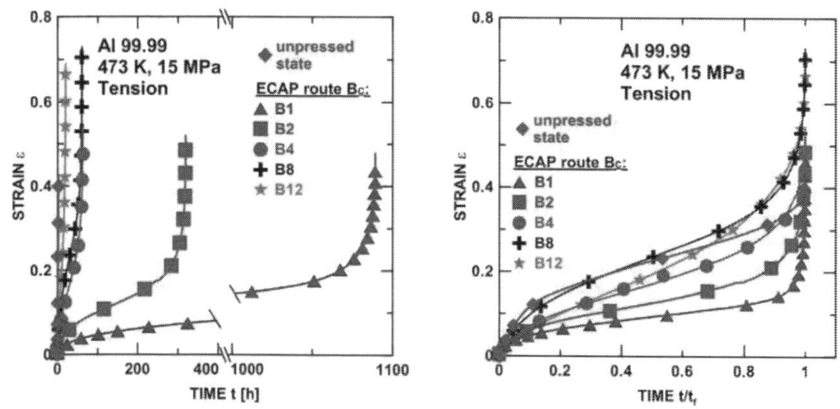

Figure 32. Creep curves of pure aluminium for unpressed state and various number of ECAP passes (creep in tension up to fracture): (a) standard creep curves, (b) creep strain ε vs. time t/t_f.

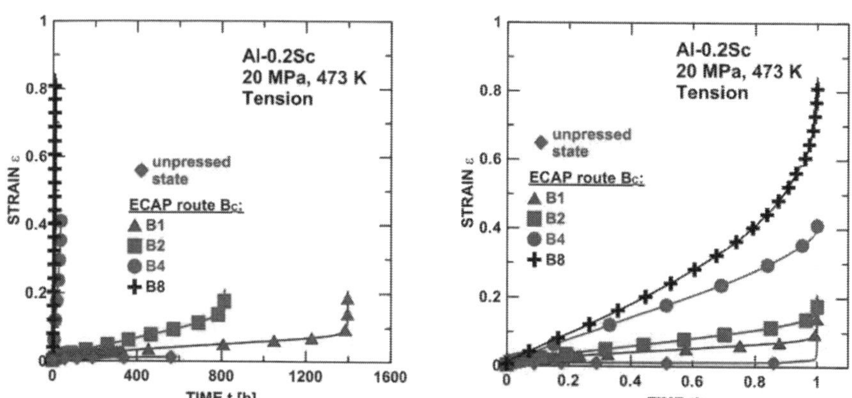

Figure 33. Creep curve of an Al-0.2wt.%Sc alloy for unpressed state and various number of ECAP passes (creep in tension up to fracture): (a) standard creep curves, (b) creep strain ε vs. time t/t_f.

It is apparent from Figure 34a that specimens with higher fraction of high-angle grain boundaries exhibit higher ductility. Further approach to the problem of ductility enhancement at room temperature in ultrafine-grained materials was suggested through the introduction of a bimodal or multimodal grain size distribution [79-81]. In this ductility strategy, the ultrafine-grained matrix in the bimodal microstructure provides the high strength, while the relatively large grains of the order of micrometers contribute to the ductility. The existence of large grains may also release stress concentrations, thereby delay the early fracture of the specimens, and allow further plastic deformation to take place in the ultrafine-grained matrix. Furthermore, the investigation of copper showed that bimodal structures may increase the ductility not only during tensile testing but also during cyclic deformation [82].

Unfortunately, the results of this work have not advocated the advantages of using material with a bimodal and/or multimodal grain size distribution for obtaining at the same time good creep strength and ductility at elevated or high temperatures. As it follows by inspection of Figures 19 and 20 typical feature of the boundary structure of UFG materials is its inhomogeneity; especially at low values of ECAP passes N completely different structures are observed. The dispersion of observed grain profile areas attains quite enormous values of the coefficient of variation CV_a (between 6 to 16) in the whole range of $N \leq 12$. In such circumstances, a great variability of times to fracture and creep fracture strain is a natural consequence of short as well as long-range inhomogeneity of microstructure of tested specimens. A very pronounced improvement of creep ductility and strong evidence for a significant role of microstructure homogeneity are documented in Figure 34b. These data indicate possible relationship between the ECAP microstructure and the creep ductility of the pressed materials [48]. The better is homogeneity of pressed material the better is its creep ductility. Thus, using a non-uniform grain structure by mixing ultrafine – sized and larger grains to derive enhanced ductility [83] seems not to be beneficial at elevated and/or high temperatures.

Finally, due to a frequent use of miniaturized tensile specimens in research on UFG materials the specimen dimension and/or geometry effects could be considered for ductile behaviour of these materials. Thus, the thickness effect is mainly caused by the necking geometry and/or fracture modes, and the gauge length effect originates from the strain definition [84]. Therefore, no existence of a standardized protocol of creep specimen and a great variety of specimen sizes and geometries have been used by different authors, primarily depending on SPD techniques used and on the availability of material. Thus, the results creep ductile behaviour measured using such different specimens are hardly comparable.

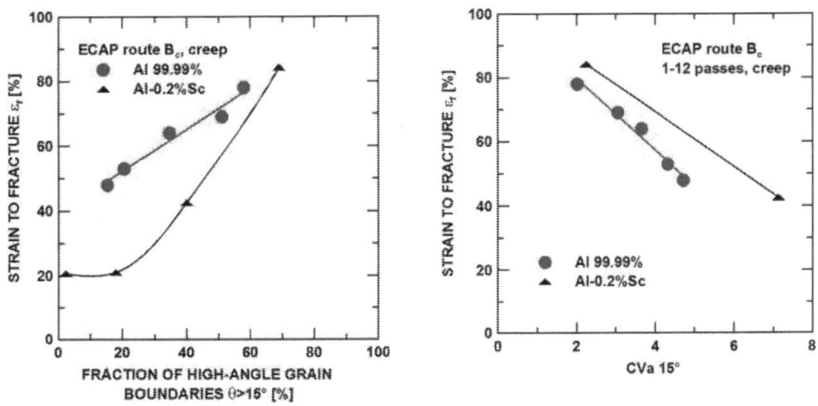

Figure 34. Dependence of the strain to fracture ε_f on (a) the fraction of high-angle grain boundaries $\theta > 15°$, and (b) on the coefficient CVa for EBSD range $\Delta \geq 15°$. Crept at 473 K and 15 MPa (Al 99.99%) or 20 MPa (Al – 0.2% Sc) – see Figures 32 and 33.

5. CONCLUSIONS

High purity (4N) aluminium, a ternary Al-3wt.%Mg-0.2wt.%Sc and a binary Al-0.2wt.%Sc alloys were processed by equal-channel angular pressing (ECAP) through 1-12 passes and examined by TEM, SEM and EBSD microscopy. The microstructural investigations reveal that from 1 to 4 ECAP passes the microstructure evolves from elongated subgrains to an essentially equiaxed array of ultrafine grains (d < 1μm) and from 4 to 12 passes there is no substantial change in average grain size. The boundary misorientation angle and the fraction of high-angle grain boundaries (HAGBs) increase rapidly up to 4 ECAP passes and at a slower rate from 4 to 12 passes.

It was found, that the creep resistance of pure aluminium is considerably increased in comparison with the coarse-grained unpressed state already after first ECAP pass. However, successive ECAP pressing lead to a noticeable decrease in the creep properties. This softening may be related to the decrease of the spacing of HAGBs at approximately constant subgrain size with increasing number of ECAP passes, resulting in the fraction of low-angle boundaries decreasing considerably. This indicates that HAGBs have lower strengthening effect under creep than low-angle ones. The softening by HAGBs may be explained in terms of the indirect effect which grain boundaries exert on the creep resistance by influencing the evolution of the dislocation microstructure in modifying the rates of generation and annihilation of dislocations. Further, a progressively increasing contribution of grain boundary sliding along HAGBs to overall creep strain can be expected as a consequence of accompanying transformation of low angle boundaries towards an equilibrium state when the number of passes increases. The results demonstrate that creep occurs in pure aluminium after processing by ECAP by the same mechanism as in conventional coarse-grained materials with intragranular dislocation glide and climb as the dominant rate-controlling deformation processes. Therefore, the higher creep resistance of pressed aluminium cannot be a consequence of any significant change in the rate controlling process.

The present results concerning Al-3wt.%Mg-0.2wt.%Sc and Al-0.2wt.%Sc alloys do not confirm a general validity of the conclusion that processing by ECAP of a coarse-grained material gives a potential for an improvement in the creep resistance. By contrast, the minimum creep rates for the pressed alloys are more rapid than the rate attained in the coarse-grained state. The creep mechanism which most probably plays the dominant role in creep of the coarse-grained alloys is the dislocation climb-bypass mechanism in the presence of elastic interaction between dislocations and coherent Al_3Sc precipitates. Strong support for making use of the same conventional creep mechanism in interpreting the creep characteristic of pressed alloy was presented. Although it seems acceptable to associate these faster creep rates with the smaller grain size after ECAP due to more intensive grain boundary sliding, there are probable further reasons strongly influencing an intragranular deformation mechanism which seems to be the rate-controlling process. First, the pressed alloys contain a high dislocation density because of the intense straining imposed during pressing. Second, an important strain contribution during creep of pressed alloys is produced by intensive mesoscale sliding of groups of grains along shear bands. Finally, the occurrence of nanopores formed during ECAP pressing at matrix/precipitate interfaces and resulting in the decohesion at coherent particle surface could have a strong effect on the creep behaviour of pressed alloys. It is important to note that the experimental results are in reasonable agreement with a theoretical model for superplastic flow that was developed earlier for coarse-grained alloys. This conclusion is important because of the many results now available demonstrating the occurrence of superplastic elongations in materials processed by ECAP and because of the new possibilities for achieving exceptional superplastic elongations in materials with ultrafine grain sizes.

6. ACKNOWLEDGEMENTS

The authors acknowledge the financial support for this work by the Czech Science Foundation under the Grant No. P108/11/2260. This work was realized in CEITEC – Central European Institute of Technology with research infrastructure supported by the project CZ.1.05/1.1.00/02.0068 financed from European Regional Development Fund. We would like to cordially thank late Dr. Ivan Saxl of Mathematical Institute of the Academy of Sciences of the Czech Republic for his useful advices concerning the correct applications and interpretations of strereological methods for quantitative characterization of microstructure.

REFERENCES

1. R. Z. Valiev, 2011 Producing Bulk Nanostructured Metals and Alloys by Severe Plastic Deformation (SPD) In: Whang S H, editor. Nanostructured Metals and Alloys: Processing, Microstructure, Mechanical Properties and Applications Cambridge Woodhead Publishing Ltd. 3 39

2. U. Erb, G. Palumbo, J. L. McCrea, 2011 The Processing of Bulk Nanocrystalline Metals and Alloys by Electrodeposition In: Whang S H, editor. Nanostructured Metals and Alloys: Processing, Microstructure, Mechanical Properties and Applications Cambridge Woodhead Publishing Ltd. 118 177

3. R. Z. Valiev, R. K. Islamgaliev, I. V. Alexandrov, 2000 Bulk Nanostructured Materials from Severe Plastic Deformation Progr Mater Sci 45 103 189

4. R. Z. Valiev, T. G. Langdon, 2006 Principles of Equal-Channel Angular Pressing as a Processing Tool for Grain Refinement Progr Mater Sci 51 881 981

5. Y. T. Zhu, R. Z. Valiev, T. G. Langdon, N. Tsuji, K. Lu, 2010 Processing of Nanostructured Metals and Alloys via Plastic Deformation MRS Bulletin 35 977 981

6. R. Z. Valiev, T. G. Langdon, 2011 Achieving Exceptional Grain Refinement Through Severe Plastic Deformation: New Approaches for Improving the Processing Technology Metall Mater Trans 42 2942 2951

7. K. Nakashima, Z. Horita, M. Nemoto, T. G. Langdon, 2000 Development of a Multi-Pass Facility for Equal-Channel Angular Pressing to High Total Strains Mater Sci Eng A 281 82 87

8. M. Furukawa, Y. Iwahashi, Z. Horita, M. Nemoto, T. G. Langdon, 1998 The Shearing Characteristics Associated with Equal-Channel Angular Pressing Mater Sci Eng A 257 328 332

9. Y. Iwahashi, Z. Horita, M. Nemoto, T. G. Langdon, 1998 The Process of Grain Refinement in Equal-Channel Angular Pressing Acta Mater 46 3317 3331

10. T. R. Mc Nelley, D. L. Swisher, Z. Horita, T. G. Langdon, 2002 Influence of Processing Route on Microstructure and Grain Boundary Development During Equal-Channel Angular Pressing of Pure Aluminium In: Zhu Y T et al., Ultrafine Grained Materials II Warrendale TMS 15 24

11. V. Sklenicka, J. Dvorak, M. Svoboda, P. Kral, B. Vlach, 2005 Effect of Processing Route on Microstructure and Mechanical Behaviour of Ultrafine-Grained Metals Processed by Severe Plastic Deformation Mater Sci Forum 482 83 88

12. J. Dvorak, V. Sklenicka, Z. Horita, 2008 Microstructural Evolution and Mechanical Properties of High Purity Aluminium Processed by Equal-Channel Angular Pressing Mater Trans 49 15 19

13. V. I. Betekhtin, A. G. Kadomtsev, V. Sklenicka, I. Saxl, 2007 Nanoporosity of Fine-Crystalline Aluminium and an Aluminium-Based Alloy Phys Solid State 49 1787 1790

14. V. I. Betekhtin, A. G. Kadomtsev, P. Kral, J. Dvorak, M. Svoboda, I. Saxl, V. Sklenicka, 2007 Significance of Microdefects Induced by ECAP in Aluminium, Al-0.2%Sc Alloy and Copper Mater Sci Forum 567-568 93 96

15. V. I. Betekhtin, V. Sklenicka, I. Saxl, B. K. Kardashev, A. G. Kadomtsev, M. V. Narykova, 2010 Influence of the Number of Passes under Equal-Channel Angular Pressing on the Elastic-Plastic Properties, Durability, and Defect Structure of the Al-0.2wt%Sc Alloy Phys Solid State 52 1517 1523

16. V. I. Betekhtin, A. G. Kadomtsev, V. Sklenicka, M. V. Narykova, 2011 Effect of Hydrostatic Pressure on Defect Structure and Durability of Ultrafine-Grained Aluminium Tech Phys Letters 37 977 979

17. V. I. Betekhtin, A. G. Kadomtsev, B. K. Kardashev, 2006 Elasticity and Anelasticity of Microcrystalline Aluminum Samples Having Various Deformation and Thermal History Phys Solid State 48 1506 1512

18. V. I. Betekhtin, E. D. Tabachnikova, A. G. Kadomtsev, M. V. Narykova, M. V. Lapovok, 2011 Effect of Counterpressure During Equal-Channel Angular Pressing on Nanoporosity Formation in Ultrafine-Grained Copper Tech Phys Letters 37 767 768

19. Yu. R. Kolobov, J. V. Ratochka, 2005 Grain Boundary Diffusion and Plasticity/Superplasticity of Polycrystalline and Nanostructured Metals and Alloys Mater Sci Eng A 410- 411 468 471

20. F. J. Humphreys, 2001 Grain and Subgrain Characterization by Electron Backscatter Diffraction J Mater Sci 36 3833 3854

21. l. Ilucova, I. Saxl, M. Svoboda, V. Sklenicka, P. Kral, 2007 Structure of ECAP Aluminium after Different Number of Passes Image Anal Stereol 26 37 43

22. Saxl, V. Sklenicka, L. Ilucova, M. Svoboda, J. Dvorak, P. Kral, 2009 The Link Between Microstructure and Creep in Aluminium Processed by Equal-Channel Angular Pressing Mater Sci Eng A 503 82 85

23. Saxl, A. Kalouskova, L. Ilucova, V. Sklenicka, 2009 Grain and Subgrain Boundaries in Ultrafine-Grained Materials Mater Characterization 60 1163 1167

24. N. Gao, S. Wang, H. S. Ubhi, M. A. Starink, 2005 A Comparison of Grain Size Determination by Light Microscopy and EBSD Analysis J Mater Sci 40 4971 4974

25. V. Sklenicka, J. Dvorak, M. Svoboda, 2004 Creep Behaviour of Pure Aluminium Processed by Equal-Channel Angular Pressing In: Zehetbauer M J, Valiev R Z, editors. Nanomaterials by Severe Plastic Deformation Weinham Wiley VCH 200 206

26. V. Sklenicka, J. Dvorak, M. Svoboda, 2004 Creep in Ultrafine-Grained Aluminium Mater Sci Eng A 387-389 696 701

27. V. Sklenicka, J. Dvorak, P. Kral, Z. Stonawska, M. Svoboda, 2005 Creep Processes in Pure Aluminium Processed by Equal-Channel Angular Pressing Mater Sci Eng A 410-411 408 412

28. K. Venkateswarlu, V. Rajinikanth, A. K. Ray, C. Xu, T. G. Langdon, 2010 The Characteristics of Aluminum-Scandium Alloys Processed by ECAP Mater Sci Eng A 527 1448 1452

29. J. Royset, N. Ryum, 2005 Kinetics and Mechanisms of Precipitation in an Al-0.2wt.%Sc Alloy Mater Sci Eng A 396 409 422

30. M. Furukawa, A. Utsunomiya, K. Matsubara, Z. Horita, T. G. Langdon, 2001 Influence of Magnesium on Grain Refinement and Ductility in a Dilute Al-Sc Alloy Acta Mater 49 3829 3838

31. E. Marquis, D. N. Seidman, D. C. Dunand, 2003 Effect of Mg Addition on the Creep and Yield Behaviour of an Al-Sc Alloy Acta Mater 51 4751 4760

32. S. Komura, Z. Horita, M. Furukawa, M. Nemoto, T. G. Langdon, 2001 An Evaluation of the Flow Behavior During High Strain Rate Superplasticity in an Al-Mg-Sc Alloy Metall Mater Trans A 32 707 716

33. V. Sklenicka, J. Dvorak, M. Svoboda, P. Kral, M. Kvapilova, Z. Horita, 2006 Compressive Creep in an Al-3%Mg-0.2%Sc Alloy Processed by Equal-Channel Angular Pressing In: Zhu Y, Langdon T G, Horita Z, Zehetbauer M J, Semiatin S L, Lowe T C, editors. Ultrafine Grained Materials IV Warrendale TMS 459 464

34. V. Sklenicka, J. Dvorak, M. Kvapilova, M. Svoboda, P. Kral, I. Saxl, Z. Horita, 2007 Effect of Equal-Channel Angular Pressing (ECAP) on Creep in Aluminium Alloys Mater Sci Forum 539-543 2904 2909

35. P. Kral, J. Dvorak, V. Sklenicka, 2008 Microstructural Evolution and Creep of an Al-0.2wt.%Sc Alloy After Equal-Channel Angular Pressing Mater Sci Forum 584-586 846 851

36. M. Kawasaki, V. Sklenicka, T. G. Langdon, 2010 An Evaluation of Creep Behavior in Ultrafine-Grained Aluminum Alloys Processed by ECAP J Mater Sci 45 271 274

37. M. Kawasaki, V. Sklenicka, T. G. Langdon, 2011 Creep Behavior of Metals Processed by Equal-Channel Angular Pressing Kovove Mater 49 75 83

38. E. A. Marquis, D. N. Seidman, 2001 Nanoscale Structural Evolution of Al3Sc Precipitates in Al(Sc) Alloys Acta Mater 49 1909 1919

39. V. Sklenicka, P. Kral, J. Dvorak, M. Kvapilova, M. Kawasaki, T. G. Langdon, 2011 Effect of Equal-Channel Angular Pressing on the Creep Resistance of Precipitation-Strengthened Alloys Mater Sci Forum 667-669 897 902

40. H. Hahn, P. Mondal, K. Podmanabhan, 1997 Plastic Deformation of Nanocrystalline Materials NanoStruc Mater 9 603 607

41. Vinogradov, S. Hashimoto, V. Patlan, K. Kitagawa, 2001 Atomic Force Microscopic Study on Surface Morphology of Ultra-Fine Grained Materials After Tensile Testing Mater Sci Eng A 319-321 862 866

42. Y. Huang, T. G. Langdon, 2003 Using Atomic Force Microscopy to Evaluate the Development of Mesoscopic Shear Planes in Materials Processed by Severe Plastic Deformation Mater Sci Eng A 358 114 121

43. V. Sklenicka, J. Dvorak, P. Kral, M. Svoboda, I. Saxl, 2009 Some Factors Affecting the Creep Behaviour of Metallic Materials Processed by Equal-Channel Angular Pressing Int J Mat Res 100 762 766

44. J. Prochazka, P. Ponizil, I. Saxl, 2008 Grain Size Estimation in Anisotropic Materials Mater Sci Forum 567- 568 285 288

45. Saxl, V. Sklenicka, 2009 Interpretation of Profile and Intercept Counts in Microstructure Characterization Mater Sci Forum 604-605 403 410

46. P. Kral, J. Dvorak, M. Kvapilova, V. Benes, P. Ponizil, O. Sedivy, V. Sklenicka, 2011 Quantitative Characterization of Microstructure in Copper Processed by Equal-Channel Angular Pressing Mater Sci Forum 667-660 235 240

47. Saxl, K. Sulleiova, P. Ponizil, 2001 Simulating Grain Size Estimation Kovove Mater 39 396 409

48. V. Sklenicka, P. Kral, L. Ilucova, I. Saxl, J. Dvorak, M. Svoboda, 2006 Inhomogeneity of Microstructure and Creep of ECAP Aluminium Mater Sci Forum 503-504 245 250

49. V. Sklenicka, J. Dvorak, P. Kral, M. Svoboda, M. Kvapilova, T. G. Langdon, 2012 Creep Ductility of Ultrafine-Grained Metallic Materials Submited to Mat Sci Eng A

50. R. Kapoor, N. Kumar, R. S. Mishra, C. S. Huskamp, K. K. Sankaran, 2010 Influence of Fraction of High Angle Boundaries on the Mechanical Behavior of an Ultrafine Grained Al-Mg Alloy Mater Sci Eng A 527 5246 5254

51. M. P. Phaniraj, M. J. N. V. Prasad, A. H. Chokshi, 2007 Grain-Size Distribution Effects in Plastic Flow and Failure Mater Sci Eng A 463 231 237

52. H. Hasegawa, S. Komura, A. Utsunomiya, Z. Horita, M. Furukawa, M. Nemoto, T. G. Langdon, 1999 Thermal Stability of Ultrafine-Grained Aluminium in the Presence of Mg and Zr Additions Mater Sci Eng A 265 188 196

53. H. Mughrabi, H. W. Höppel, M. Kautz, R. Z. Valiev, 2003 Annealing Treatments to Enhance Thermal and Mechanical Stability of Ultrafine-

Grained Metals Produced by Severe Plastic Deformation Z Metallkde 94 1079 1083

54. M. Kawasaki, I. J. Beyerline, S. C. Vogel, T. G. Langdon, 2008 Characterization of Creep Properties and Creep Textures in Pure Aluminum Processed by Equal-Channel Angular Pressing Acta Mater 56 2307 2317

55. B. Wilshire, C. Palmer, 2002 Deformation Processes During Creep of Pure Aluminium In: Mishra R S, Earthman J C, Raj S V, editors. Creep Deformation. Fundamentals and Applications Warrendale TMS 51 60

56. B. Wilshire, P. J. Scharning, 2008 Creep and Creep Fracture of Commercial Aluminium Alloys J Mater Sci 43 3992 4000

57. V. Sklenicka, J. Dvorak, M. Svoboda, 2004 Influence of Processing Route on Creep of Ultrafine Grained Aluminium Prepared by ECAP In: Zhu Y T, Langdon T G, Valiev R Z, Semiatin S L, Shin D H, Lowe T C, editors. Ultrafine Grained Materials III Warrendale TMS 647 652

58. V. Sklenicka, J. Cadek, 1970 Determination of Strain Component Caused by Grain Boundary Sliding Z Metallkde 61 575 580

59. J. Cadek, 1988 Creep in Metallic Materials Amsterdam Elsevier Science Publ. 372

60. T. G. Langdon, 2006 Grain Boundary Sliding Revisited. Developments in Sliding over Four Decades J Mater Sci 41 597 609

61. Zhilyaev, A. Pshenichnyuk, 2011 Superplasticity and Grain Boundaries in Ultrafine-Grained Materials Cambridge Woodhead Publishing 312

62. S. V. Divinski, G. Reglitz, H. Rosner, Y. Estrin, G. Wilde, 2011 Ultra-Fast Diffusion Channels in Pure Ni Severly Deformed by Equal-Channel Angular Pressing Acta Mater 59 1974 1985

63. F. R. N. Nabarro, 1948 Deformation of Crystals by the Motions of Single Ions In: Rep. Conf. Strength Solids The Physical Society London 75

64. C. Herring, 1950 Diffusional Viscosity of a Polycrystalline Materials J Appl Phys 21 437

65. R. L. Coble, 1963 A Model for Boundary Diffusion Controlled Creep in Polycrystalline Materials J Appl Phys 31 1679

66. F. A. Mohamed, T. G. Langdon, 1974 Deformation Mechanism Maps Based on Grain-Size Metall Trans 5 2239 2345

67. X. H. Zeng, Y. J. Li, W. Blum, 2004 On Coble Creep in Ultrafine-Grained Cu Phys Stat Sol (a) 201 14 R114 117

68. P. Yavari, F. A. Mohamed, T. G. Langdon, 1981 Creep and Substructure Formation in an Al-5Percent Mg Solid Solution Alloy Acta Metall 29 1495 1507

69. P. Yavari, T. G. Langdon, 1982 An Examination of the Breakdown in Creep by Viscous Glide in Solid Solution Alloys at High Stress Levels Acta Metall 30 2181 2196

70. T. R. Mc Nelley, D. J. Michel, A. Salama, 1989 The Mg-Concentration Dependence of the Strength of AlMg Alloys During Glide-Controlled Deformation Scripta Met 23 1657 1662

71. C. B. Fuller, D. N. Seidman, D. C. Dunand, 1999 Creep Properties of Coarse-Grained Al(Sc) Alloys at 300°C Scripta Mater 40 691 696

72. D. N. Seidman, E. A. Marquis, D. C. Dunand, 2002 Precipitation Strengthening at Ambient and Elevated Temperatures of Heat-Treatable Al(Sc) Alloys Acta Mater 50 4021 4035

73. J. C. Gibeling, W. D. Nix, 1980 The Description of Elevated Temperature Deformation in Terms of Threshold Stresses and Back Stresses: A Review Mater Sci Eng 45 123 135

74. E. A. Marquis, D. C. Dunand, 2002 Model for Creep Threshold Stress in Precipitation-Strengthened Alloys with Coherent Particles Scripta Mat 47 503 508

75. J. Cadek, S. J. Zhu, K. Milicka, 1998 Threshold Creep Behaviour of Aluminium Dispersion Strengthened by Fine Alumina Particles Mater Sci Eng A 252 1 5

76. W. Blum, P. Eisenlohr, V. Sklenicka, 2009 Creep Behavior of Bulk Nanostructured Materials- time dependent deformation and deformation kinetics In: Zehetbauer M J, Zhu T Z, editors. Bulk Nanostructured Materials Weinhaim Wiley-VCH Verlag GmBH & Co. 519 538

77. Y. R. Kolobov, G. P. Grabovetskaya, M. B. Ivanov, A. P. Zhilyaev, R. Z. Valiev, 2001 Grain Boundary Diffusion Characteristics of Nanostructured Nickel Scripta Mater 44 873 878

78. Z. B. Wang, K. Lu, G. Wilde, S. V. Divinski, 2011 Effect of Grain Growth on Interface Diffusion in Nanostructured Cu Scripta Mater 64 1055 1058

79. Y. Wang, M. Chen, F. Zhou, E. Ma, 2002 High Tensile Ductility in a Nanostructured Metal Nature 419 912 914

80. M. Kawasaki, Z. Horita, T. G. Langdon, 2009 Microstructural Evolution in High Purity Aluminum Processed by ECAP Mater Sci Eng A 524 143 150

81. N. Llorca-Isern, T. Grosdidier, J. M. Cabrera, 2010 Enhancing Ductility of ECAP Pressed Metals Mater Sci Forum 654-656 1219 1222

82. M. Korn, R. Lapovok, A. Böhner, H. W. Höppel, H. Mughrabi, 2011 Bimodal Grain Size Distributions in UFG Materials Produced by SPD-Their Evolution and Effect on the Fatigue and Monotonic Strength Properties Kovove Mater 49 51 53

83. E. Ma, 2003 Instabilities and Ductility of Nanocrystalline and Ultrafine-Grained Metals Scripta Mater 49 663 668

84. Y. H. Zhao, Y. Z. Guo, Q. Wie, A. M. Dangelewitz, C. Xu, Y. T. Zhu, T. G. Langdon, Y. Z. Zhou, E. J. Lavernia, 2008 Influence of Specimen Dimensions on the Tensile Behaviour of Ultrafine-Grained Cu Scripta Mater 59 627 630

CHAPTER 3

Fabrication of Corrosion Resistance Micro-Nanostructured Superhydrophobic Anodized Aluminum in a One-Step Electrodeposition Process

Ying Huang, Dilip K. Sarkar * and X.-Grant Chen

Centre Universitaire de Recherche sur l'Aluminium (CURAL), Université du Québec à Chicoutimi, 555 Boulevard de l'Université, Chicoutimi, QC G7H 2B1, Canada

ABSTRACT

The formation of low surface energy hybrid organic-inorganic micro-nanostructured zinc stearate electrodeposit transformed the anodic aluminum oxide (AAO) surface to superhydrophobic, having a water contact angle of 160°. The corrosion current densities of the anodized and aluminum alloy surfaces are found to be 200 and 400 nA/cm^2, respectively. In comparison, superhydrophobic anodic aluminum oxide (SHAAO) shows a much lower value of 88 nA/cm^2. Similarly, the charge transfer resistance, R_{ct}, measured by electrochemical impedance spectroscopy shows that the SHAAO substrate was found to be 200-times larger than the as-received aluminum alloy substrate. These results proved that the superhydrophobic surfaces created on the anodized surface significantly improved the corrosion resistance property of the aluminum alloy.

KEYWORDS

superhydrophobic aluminum; corrosion; anodized aluminum oxide (AAO); organic-inorganic; micro-nanostructure; zinc stearate (ZnSA); potentiodynamic polarization; electrochemical impedance spectroscopy (EIS)

1. INTRODUCTION

Aluminum (Al) and its alloys are naturally-abundant engineering materials with extensive applications in daily life. In recent years, nanoporous anodic aluminum oxide (AAO) prepared by electrochemical anodization has found a multitude of applications, such as catalysis [1], drug delivery [2], biosensing [3], template synthesis [4], molecular and ion separation [5], corrosion resistance [6], and so forth. The formation of AAO on the aluminum alloy surface would act as the corrosion barrier. The formation of AAO was limited to a certain extent due to its hydrophilic behavior. Therefore, it is necessary to transform the AAO to be superhydrophobic in order to improve the corrosion resistance properties.

Superhydrophobicity, exhibiting an excellent water-repellent property, is characterized by a contact angle above 150°. Creating a rough surface, as well as reducing the surface energy is attributed to the modification of superhydrophobicity [7]. In the last few decades, a large effort has been devoted to the realization of superhydrophobic surfaces, due to their applications in biology [8], anti-corrosion [9,10,11,12], anti-icing [13], self-cleaning [14], *etc.*

Recent publications show that the anodized surface can be made superhydrophobic by passivation with organic molecules [15,16,17,18]. In the study of Liu *et al.*, the superhydrophobic anodized surfaces were fabricated by polypropylene (PP) coating after anodizing [17]. Li *et al.*, used a very complex process to engineer superhydrophobic anodized aluminum alloy surfaces [16]. After anodizing with sulfuric acid, the anodized sample was firstly immersed in the mixing solution containing $M(NO_3)_2$ salt (M = Mg, Co, Ni and Zn) and NH_4NO_3, followed by immersion in (Heptadecafluoro-1,1,2,2-tetradecyl)trimethoxysilane (n-$CF_3(CF_2)_7CH_2CH_2Si(OC_2H_5)_3$). Vengatesh and Kulandainathan fabricated superhydrophobic anodic aluminum oxide (SHAAO) surfaces by passivation with organic molecules and show these surfaces having corrosion resistance properties indicated by potentiodynamic polarization [15]. Despite this, the fabrication, as well as corrosion resistance properties of the electrodeposited superhydrophobic surfaces on AAO have not been shown in the literature.

In this work, we have described the method to prepare the superhydrophobic surfaces on AAO by the electrodeposition process and describe their corrosion resistance properties, both using potentiodynamic polarization and electrochemical impedance spectroscopy (EIS).

2. EXPERIMENTAL PROCEDURE

After pretreatment with 0.01 M NaOH at 55 °C for 3 min, the AA6061 aluminum alloy substrates were anodized using 3 vol. % H_3PO_4 aqueous solution at 10 °C at 0.01 A/cm^2 (Ametek Sorensen DCS 100-12E, Chicoutimi, QC, Canada) for 2 h. The superhydrophobic anodic aluminum oxide (SHAAO) surface was prepared by the electrodeposition process in an ethanolic solution containing 0.01 M stearic acid (SA) and 0.01 M zinc nitrate ($Zn(NO_3)_2$) by applying 20 V (Ametek Sorensen DCS 100-12E) for 10 min. Microstructural examination was conducted using a scanning electron microscope (SEM, JEOL JSM-6480 LV, Chicoutimi, QC, Canada). Surface roughness was measured using an optical profilometer. The X-ray diffraction (XRD) analyses of the samples were carried out using a Bruker D8 Discover system (Chicoutimi, QC, Canada). The chemical composition of the samples was analyzed by means of Fourier transform infrared spectroscopy (FTIR, Perkins Elmer Spectrum One, Chicoutimi, QC, Canada) and an energy dispersive X-ray spectrometer (EDX, JEOL JSM-6480 LV, Chicoutimi, QC, Canada). Wetting characteristics of sample surfaces were evaluated by measuring static contact angles (CA) using a first ten Angstrom contact angle goniometer at five positions on each substrate using a 10-µL deionized water drop. In the case of rolling-off surfaces, the contact angle was measured by holding the water drop between the needle and the surface, as presented in the inset image of Figure 1c. Electrochemical experiments were performed using a PGZ100 potentiostat and a 300-cm^3-EG & G PAR flat cell (London Scientific, London, ON, Canada), equipped with a standard three-electrode system with an Ag/AgCl reference electrode, a platinum (Pt) mesh as the counter electrode (CE) and the sample as the working electrode (WE). Before the test, the open circuit potential (OCP) was monitored for more than 20 h for stabilization by immersing the sample surface in 3 wt. % NaCl aqueous solution. The electrochemical impedance spectroscopy (EIS) was tested in the frequency range between 10 MHz and 100 kHz with a sine-wave amplitude of 10 mV. For the potentiodynamic polarization experiments, the potential was scanned from −250 mV to +1000 mV with respect to the OCP voltage at a scan rate of 2 mV/s.

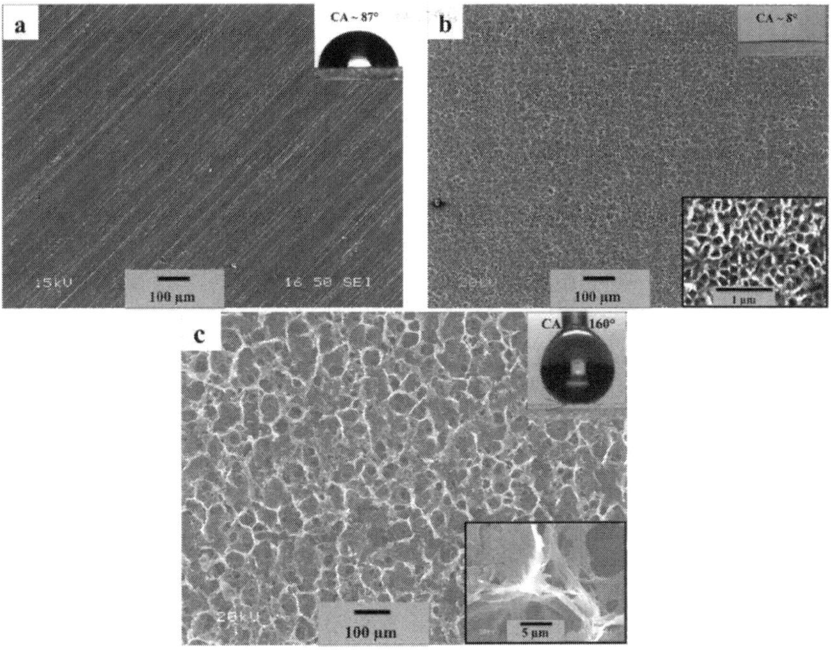

Figure 1. Secondary electron SEM micrographs showing the top surface of (**a**) the as-received aluminum alloy; (**b**) the anodized and (**c**) the electrodeposited anodized substrates. The inset top-right images present the water drop on the surfaces, and the bottom-right ones show magnified microstructures.

3. RESULTS AND DISCUSSION

SEM images in Figure 1 reveal the evolution of the different morphologies of: (a) the as-received the aluminum alloy surface; (b) the anodized (AAO) surface; and (c) the electrodeposited hybrid organic-inorganic anodized surface using the ethanolic electrolytic solution containing stearic acid (SA) and zinc nitrate ($Zn(NO_3)_2$) at 20 V DC. The as-received aluminum alloy surface was characterized by parallel lines resulting from the rolling process, corresponding to a surface roughness of 0.45 ± 0.03 μm and a contact angle of 87° ± 3° (Figure 1a).

After anodizing in phosphoric acid for a duration of 2 h, uniformly-distributed nanopores with an average diameter of approximately 100 nm and an inter-pore distance of ~137 nm are observed on the surface of the anodized substrate (Figure 1b), with a roughness of 0.68 ± 0.02 μm, while the contact angle was measured to be 8° ± 1°, indicating a superhydrophilic property. It is however evident from Figure 1c that the electrodeposition process resulted in the appearance of a porous network microstructure on the anodized substrate. This substrate was built by nanofibre clusters connected with each other, as presented in the inset of Figure 1c. It can be also observed that these micro-nanoporous structures are distributed uniformly on the anodized surface, resulting in a micro-nanorough surface having a roughness of 6.85 ± 1.02 μm. Interestingly, this surface shows superhydrophobic properties with a contact angle of 160° ± 1° having a contact angle hysteresis of 2° ± 1°.

Low angle X-ray diffraction (XRD), energy dispersive X-ray spectroscopy (EDX) and Fourier transform infrared spectroscopy (FTIR) have been carried out to determine the composition of the electrodeposited micro-nanostructure thin films on the anodized substrate, as shown in Figure 2. Figure 2a(a3) shows four distinct peaks at 4.2°, 6.26°, 8.3° and 10.4°, which correspond to zinc stearate ($(CH_3(CH_2)_{16}COO)_2Zn$) (abbreviated as ZnSA,). The possible mechanism of the formation of ZnSA has been presented as follows:

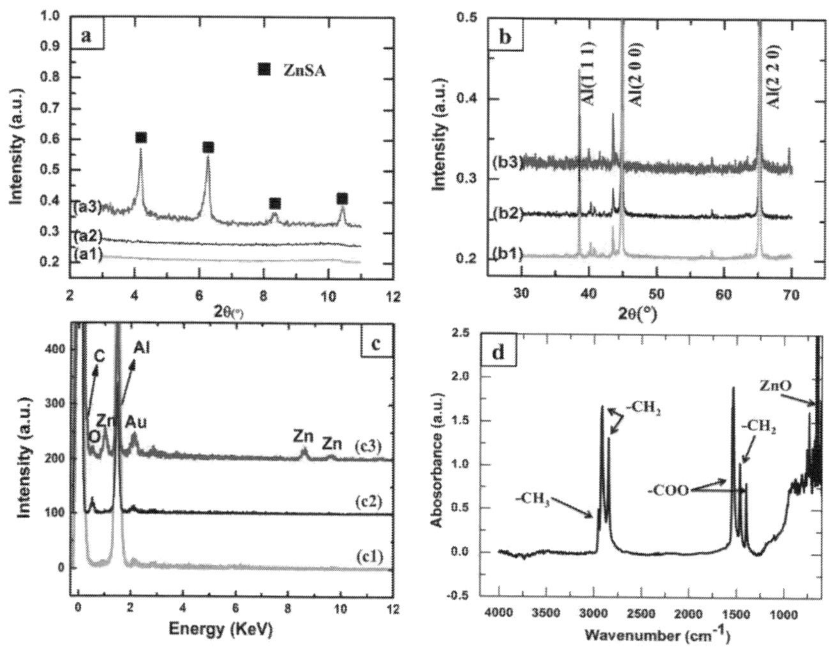

Figure 2. (**a**) Low angle XRD patterns of (**a1**) as-received aluminum alloy, (**a2**) AAO, (**a3**) electrodeposited anodized substrate; (**b**) high angle XRD patterns of (**b1**) as-received aluminum alloy, (**b2**) AAO and (**b3**) electrodeposited anodized substrate; (**c**) EDX spectra of (**c1**) the as-received aluminum alloy, (**c2**) anodized and (**c3**) electrodeposited anodized substrate (ZnSA = zinc stearate); (**d**) FTIR spectrum of the electrodeposited anodized substrate.

$$2CH_3(CH_2)_{16}COOH + Zn^{2+} \xrightarrow{20\,V} (CH_3(CH_2)_{16}COO)_2Zn + 2H^+ \quad (1)$$

This reaction mechanism is very similar to that mentioned by Liu *et al.* for the electrodeposition of cerium myristate by electrodeposition on magnesium substrates to obtain superhydrophobicity [19]. When the DC voltage was applied on the electrodes, the Zn^{2+} ions close to the cathodic electrode reacted with SA, forming ZnSA and H^+ ions, as presented in Equation (1). Meanwhile, some of the H^+ ions obtained an electron and formed H_2 gas. It is noted that the as-received aluminum alloy and anodized substrates of Figure 2a(a1,a2) do not show any characteristic peaks. Furthermore, in Figure 2b, it shows the peaks at 38.47°, 44.72° and 65.1°, respectively, which are in good agreement with the characteristic peaks of Al(111), (200) and (220). This arises from the substrate of the aluminum alloy.

EDX analysis (Figure 2c) revealed that the chemical composition of the anodized surface consisted of O and Al, resulting from the formation of aluminum oxide (Al_2O_3), whereas only the Al peak was seen in the spectrum of the as-received aluminum alloy surface. However, C, Zn, O and Al are observed in the spectrum of electrodeposited anodized surface, confirming the formation of ZnSA complementary with the XRD pattern of the electrodeposited anodized substrate. It is worth mentioning that the Au peaks appearing in Figure 2c are due to the thin layer of gold on the electrodeposited thin films for improving the resolution by eliminating the charging effect of non-conducting samples during EDX analyses.

In the FTIR spectrum of the electrodeposited hybrid organic-inorganic anodized substrate (Figure 2d), the appearance of the –CH group ($-CH_3$ at 2962 cm^{-1}, as well as $-CH_2$ at 1459 cm^{-1}, 2850 cm^{-1} and 2919 cm^{-1}), as well as –COO at 1395 and 1550 cm^{-1} indicated the formation of ZnSA on the surface [20], which is in good agreement with the XRD and EDX results. These results support the formation of low surface energy methylated ($-CH_3$ and $-CH_2$) components on the electrodeposited anodized surface and make it superhydrophobic. In addition, the ZnO peaks at 560 cm^{-1} may come from the bonding of –COOZn, which further verifies the reaction production of ZnSA formed on the modified surface in a one-step electrodeposition process.

Figure 3 illustrates the variation of open circuit potential (OCP) on the superhydrophobic AAO (SHAAO) surface. The OCP value shifted from −177 to −708 mV with an average of −281 ± 87 mV into a 5-h immersion time in the salt solution. It is quite unstable during this period. From 5 to 16 h of immersion time, the OCP value varied from −533 to −686 mV with an average of −610 ± 26.8 mV. With the prolongation of immersion time to 20 h, the OCP value varied from −708 to −730 mV, with an average of −717 ± 2.8 mV. It can be seen that the fluctuation of OCP voltage reduces with time and stabilized nearly after 20 h. In contrast, the surface of the aluminum alloy substrate gets stabilized within 30 min of immersion time in the salt solution. The OCP fluctuations of the SHAAO substrate may be due to the poor wetting, as well as protective properties of the SHAAO surface with the salt solution. Therefore, the EIS and polarization experiments were performed after 20 h of immersion of the superhydrophobic surface in the salt solution while monitoring the OCP continuously.

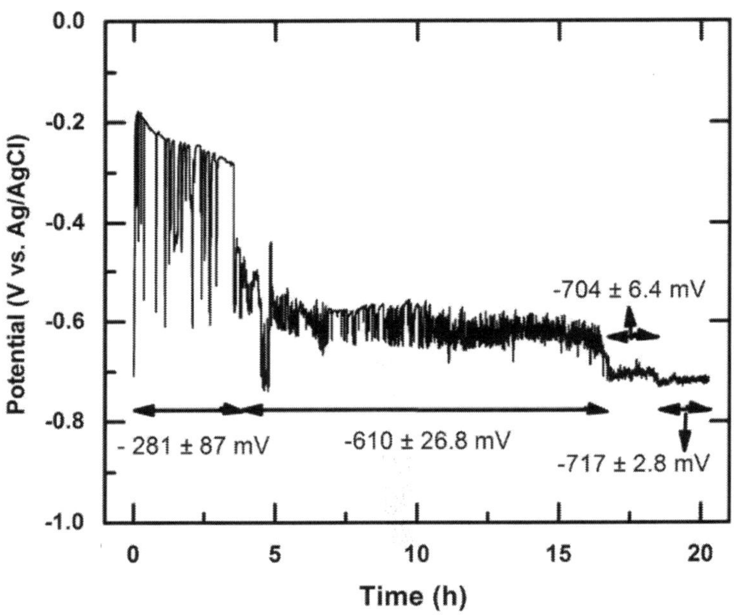

Figure 3. The variation of open circuit potential (OCP) with time on the superhydrophobic AAO (SHAAO) surface.

To evaluate the corrosion resistance performance of the fabricated anodized and SHAAO substrates, potentiodynamic polarization and electrochemical impedance spectroscopy (EIS) tests were carried out. Figure 4a shows the polarization curves of the as-received aluminum alloy, anodized and SHAAO substrates, respectively. The as-received aluminum alloy substrate exhibited a corrosion current density (I_{corr}) of 400 nA/cm^2 and a polarization resistance (R_p) of 50 kΩ·cm^2. The $R_p$value was calculated by the Stern-Geary equation, as shown in Equation (2).

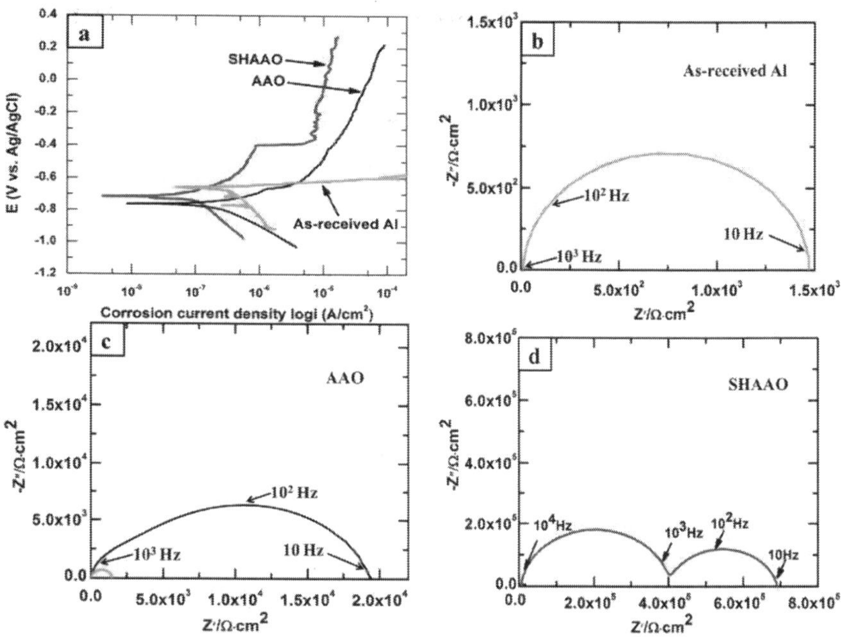

Figure 4. (**a**) Polarization curves of the as-received aluminum alloy, AAO and SHAAO substrates in 3 wt. % NaCl corrosive solution; (**b–d**) the Nyquist plots of (**b**) as-received aluminum alloy; (**c**) AAO (as-received aluminum alloy (a small semicircle in green close to the origin) also shown for comparison) and (**d**) SHAAO substrate (as-received aluminum alloy and AAO also shown inside for comparison, very small semicircle, nearly visible, close to the origin).

$$Rp = \frac{\beta_a \beta_c}{2.3 I_{corr}(\beta_a + \beta_c)} \tag{2}$$

When the aluminum alloy substrate was anodized, the I_{corr} was reduced to 200 nA/cm^2, and R_p increased to 87 kΩ·cm^2, indicating an improved corrosion resistance compared to the as-received substrate. This was due to a barrier layer formed on the anodized surface. On the other hand, the I_{corr} of the hybrid organic-inorganic SHAAO substrate decreased remarkably to 88 nA/cm^2, while the R_p increased up to 441 kΩ·cm^2. This shows an even better anti-corrosion performance of the SHAAO substrate relative to the anodized substrate, likely attributed to the superhydrophobic ZnSA coating formed on the surface. It is noticed that the current density of the SHAAO substrate was increased sharply at around −0.4 V *vs.*Ag/AgCl, which might have been related to the dissolution of the superhydrophobic film with the prolongation of corrosion time (more than 20 h). It is also found that the current density of the SHAAO substrate at −0.4 V to 0.2 V *vs.* Ag/AgCl was parallel to, but smaller than, that of the anodized substrate. This might be due to the superhydrophobic material ZnSA filling the anodized pore structure. As a result, the SHAAO substrate after dissolution of the surface material still presents a good corrosion resistance property.

The I_{corr} of the electrodeposited cerium stearate superhydrophobic Mg surfaces was reported to be 142 nA/cm^2 with a 30-min immersion to perform OCP [19]. Experiments performed by Vengatesh and Kulandainathan with a 30-min immersion before polarization show that the corrosion current varies between 2 and 1050 nA/cm^2, depending on the passivated molecules [15]. Moreover, in the study of He *et al.*, a 1-h immersion time is used for the stabilization [21]. In our experiment, the samples were exposed to NaCl solution for 20 h to stabilize under OCP. This indicates that our superhydrophobic film displays a better stability and durability in the corrosion test as compared to the reported values in the literature.

Nyquist plots of the as-received aluminum alloy, anodized and hybrid organic-inorganic SHAAO substrate are presented in Figure 4b–d. It is well known that R_{ct} is the charge transfer resistance, describing the difficulty of the corrosion occurring on the substrate. R_{ct} of anodized substrate is found to be 14 kΩ·cm^2 in Figure 4c, which is about 10-times higher than that of the as-received aluminum alloy substrate (1.5 kΩ·cm^2) in Figure 4b. However, the corrosion protection of the SHAAO substrate is the combined effect of R_{ct} of 284 kΩ·cm^2 (semicircle at low frequency) along with the resistance of the superhydrophobic films of 405 kΩ·cm^2 (semicircle at high frequency) in Figure 4d. This indicates a significant enhancement of corrosion resistance, which is complementary to the result from the polarization curves, where the polarization resistance of SHAAO is higher than the anodized substrate, as well as for the Al substrate.

In the study of Liu *et al.* [19], the R_{ct} of the superhydrophobic magnesium substrate covered with cerium myristate was found to be 13 kΩ, which is comparable to our anodized substrate, but much smaller than our zinc stearate-covered SHAAO substrate [19].

4. CONCLUSIONS

A superhydrophobic anodic aluminum oxide (SHAAO) surface was prepared by the electrodeposition process using the ethanolic solution of stearic acid (SA) mixed with zinc nitrate ($Zn(NO_3)_2$) at a constant voltage of 20 V. The hybrid organic-inorganic SHAAO surface having a micro-nanoporous structure of low surface energy zinc stearate (ZnSA) exhibits a water contact angle (CA) of 160° ± 1°. The SHAAO substrate has a polarization resistance (R_p) and charge transfer resistance (R_{ct}) of 441 and 284 kΩ·cm^2, respectively, much higher than that of the as-received aluminum(R_p of 50 kΩ·cm^2 and R_{ct} of 1.46 kΩ·cm^2) and anodized aluminum substrate (R_p of 87 kΩ·cm^2 and R_{ct} of 14 kΩ·cm^2) in the corrosion. This indicates that the SHAAO substrate displays a much better corrosion resistance property as compared to the as-received aluminum alloy substrate, as well as the anodic aluminum oxide (AAO) substrate.

ACKNOWLEDGMENTS

We acknowledge the financial support provided by the Natural Science and Engineering Research Council of Canada (NSERC) and the Aluminium Research Centre—REGAL.

AUTHOR CONTRIBUTIONS

Ying Huang performed experiments, analysed data and wrote the draft of manuscript; D.K. Sarkar and X.-Grant Chen supervised the works. All authors took part in the discussion, the preparation and revision of the manuscript.

CONFLICTS OF INTEREST

The authors declare no conflict of interest.

REFERENCES

1. Dotzauer, D.M.; Dai, J.; Sun, L.; Bruening, M.L. Catalytic membranes prepared using layer-by-layer adsorption of polyelectrolyte/metal nanoparticle films in porous supports. *Nano Lett.* **2006**, *6*, 2268–2272.
2. Simovic, S.; Losic, D.; Vasilev, K. Controlled drug release from porous materials by plasma polymer deposition. *Chem. Commun.* **2010**, *46*, 1317–1319.

3. Wang, M.; Meng, G.; Huang, Q.; Xu, Q.; Chu, Z.; Zhu, C. FITC-modified PPy nanotubes embedded in nanoporous AAO membrane can detect trace PCB20 via fluorescence ratiometric measurement. *Chem. Commun.* **2011**, *47*, 3808–3810.

4. Liu, Y.; Goebl, J.; Yin, Y. Templated synthesis of nanostructured materials. *Chem. Soc. Rev.* **2013**, *42*, 2610–2653.

5. Jani, A.M.M.; Anglin, E.J.; McInnes, S.J.P.; Losic, D.; Shapter, J.G.; Voelcker, N.H. Nanoporous anodic aluminium oxide membranes with layered surface chemistry. *Chem. Commun.* **2009**, *21*, 3062–3064.

6. Bouchama, L.; Azzouz, N.; Boukmouche, N.; Chopart, J.P.; Daltin, A.L.; Bouznit, Y. Enhancing aluminum corrosion resistance by two-step anodizing process. *Surf. Coat. Technol.* **2013**, *235*, 676–684.

7. Huang, Y.; Sarkar, D.K.; Chen, X.-G. A one-step process to engineer superhydrophobic copper surfaces. *Mater. Lett.***2010**, *64*, 2722–2724.

8. Ciasca, G.; Papi, M.; Chiarpotto, M.; de Ninno, A.; Giovine, E.; Campi, G.; Gerardino, A.; de Spirito, M.; Businaro, L. Controlling the cassie-to-wenzel transition: An easy route towards the realization of tridimensional arrays of biological objects. *Nano-Micro Lett.* **2014**, *6*, 280–286.

9. Zhang, F.; Chen, S.; Dong, L.; Lei, Y.; Liu, T.; Yin, Y. Preparation of superhydrophobic films on titanium as effective corrosion barriers. *Appl. Surf. Sci.* **2011**, *257*, 2587–2591.

10. Yin, B.; Fang, L.; Tang, A.-Q.; Huang, Q.-L.; Hu, J.; Mao, J.-H.; Bai, G.; Bai, H. Novel strategy in increasing stability and corrosion resistance for super-hydrophobic coating on aluminum alloy surfaces. *Appl. Surf. Sci.* **2011**, *258*, 580–585.

11. Huang, Y.; Sarkar, D.K.; Chen, X.-G. Superhydrophobic aluminum alloy surfaces prepared by chemical etching process and their corrosion resistance properties. *Appl. Surf. Sci.* **2015**, *356*, 1012–1024.

12. Huang, Y.; Sarkar, D.K.; Gallant, D.; Chen, X.-G. Corrosion resistance properties of superhydrophobic copper surfaces fabricated by one-step electrochemical modification process. *Appl. Surf. Sci.* **2013**, *282*, 689–694.

13. Zuo, Z.; Liao, R.; Guo, C.; Yuan, Y.; Zhao, X.; Zhuang, A.; Zhang, Y. Fabrication and anti-icing property of coral-like superhydrophobic aluminum surface. *Appl. Surf. Sci.* **2015**, *331*, 132–139.

14. Watson, G.S.; Green, D.W.; Schwarzkopf, L.; Li, X.; Cribb, B.W.; Myhra, S.; Watson, J.A. A gecko skin micro/nano structure—A low

adhesion, superhydrophobic, anti-wetting, self-cleaning, biocompatible, antibacterial surface. *Acta Biomater.* **2015**, *21*, 109–122.

15. Vengatesh, P.; Kulandainathan, M.A. Hierarchically ordered self-lubricating superhydrophobic anodized aluminum surfaces with enhanced corrosion resistance. *ACS Appl. Mater. Interfaces* **2015**, *7*, 1516–1526.

16. Li, Y.; Li, S.; Zhang, Y.; Yu, M.; Liu, J. Fabrication of superhydrophobic layered double hydroxides films with different metal cations on anodized aluminum 2198 alloy. *Mater. Lett.* **2015**, *142*, 137–140.

17. Liu, W.; Luo, Y.; Sun, L.; Wu, R.; Jiang, H.; Liu, Y. Fabrication of the superhydrophobic surface on aluminum alloy by anodizing and polymeric coating. *Appl. Surf. Sci.* **2013**, *264*, 872–878.

18. Fujii, T.; Aoki, Y.; Habazaki, H. Superhydrophobic hierarchical surfaces fabricated by anodizing of oblique angle deposited Al–Nb alloy columnar films. *Appl. Surf. Sci.* **2011**, *257*, 8282–8288.

19. Liu, Q.; Kang, Z. One-step electrodeposition process to fabricate superhydrophobic surface with improved anticorrosion property on magnesium alloy. *Mater. Lett.* **2014**, *137*, 210–213.

20. Huang, Y.; Sarkar, D.K.; Chen, X.-G. Superhydrophobic nanostructured zno thin films on aluminum alloy substrates by electrophoretic deposition process. *Appl. Surf. Sci.* **2015**, *327*, 327–334.

21. He, T.; Wang, Y.; Zhang, Y.; Lv, Q.; Xu, T.; Liu, T. Super-hydrophobic surface treatment as corrosion protection for aluminum in seawater. *Corros. Sci.* **2009**, *51*, 1757–1761.

CHAPTER 4

Influence of Post Weld Heat Treatment on Strength of Three Aluminum Alloys Used in Light Poles

Craig C. Menzemer [1], Eric Hilty [2], Shane Morrison [3], Ray Minor [4] and Tirumalai S. Srivatsan [2,*]

[1] Civil Engineering and Associate Dean, College of Engineering, The University of Akron, Akron, OH 44325, USA

[2] Department of Civil Engineering, the University of Akron, Akron, OH 44325, USA

[3] Research and Development, HAPCO, Abingdon, VA 24210, USA

[4] HAPCO, Abingdon, VA 24210, USA

ABSTRACT

The conjoint influence of welding and artificial aging on mechanical properties were investigated for extrusions of aluminum alloy 6063, 6061, and 6005A. Uniaxial tensile tests were conducted on the aluminum alloys 6063-T4, 6061-T4, and 6005A-T1 in both the as-received (AR) and as-welded (AW) conditions. Tensile tests were also conducted on the AR and AW alloys, subsequent to artificial aging. The welding process used was gas metal arc (GMAW) with spray transfer using 120–220 A of current at 22 V. The artificial aging used was a precipitation heat treatment for 6 h at 182 °C (360 °F). Tensile tests revealed the welded aluminum alloys to have lower strength, both for yield and ultimate tensile strength, when compared to the as-received un-welded counterpart. The beneficial influence of post weld heat treatment (PWHT) on strength and ductility is presented and discussed in terms of current design provisions for welded aluminum light pole structures.

KEYWORDS

light poles; aluminum alloy; welding; gas metal arc welding; artificial aging; post weld heat treatment; microstructure; tensile properties

1. INTRODUCTION AND BACKGROUND

Over the last four decades, *i.e.*, since the early 1970s, structural aluminum alloys have been used in a myriad of applications, primarily because they can offer an attractive combination of strength, are light in weight, have a high strength-to-weight (σ/ρ) ratio, and, most importantly, are cost efficient [1]. Many products are being increasingly fabricated from 6XXX-series aluminum alloys due to their innate ability to be extruded into complex shapes, coupled with their receptiveness to welding and their notable resistance to environment-induced degradation or corrosion [2]. Understanding the weldability and resultant mechanical properties is important in an attempt to put these alloys to efficient use. It is uncommon for an aluminum alloy to be welded with no influence on microstructure and resultant mechanical properties, such as strength. However, precipitation heat treatment does offer the promise of minimizing the negative effects of welding on the mechanical properties of the family of 6XXX alloys. A product of considerable practical interest and significance is welded aluminum light poles [3,4].

A widely-used method for joining the alloys of aluminum is welding. A few noteworthy examples related to the commercial industry include the following: (i) fabrication of rail vehicles; (ii) marine structures; (iii) pressure vessels; (iv) automotive components; and (v) structures in the civil construction industry. A few noteworthy advantages of the welding process include the following: (a) high joint efficiencies; (b) flexibility; (c) speed; and (d) a low fabrication cost [5]. Welding involves "localized" melting of the base material; as a consequence of which, both the microstructure and resultant mechanical properties will be different from those of the base material [6]. To obtain improved properties for the welded material, component, or structure, a heat treatment is both necessary and essential [7].

Section 2.5 of the 2010 Aluminum Design Manual (ADM) provides mechanical property information for welded and, subsequently, heat treated alloys that are chosen for use in aluminum light poles [8]. Aluminum alloys 6005 and 6063 have been widely used for welded light poles. For poles manufactured from aluminum alloy 6005, welded in the T1 temper, and having a thickness less than or equal to 6.4 mm, the specifications allow the engineer to make effective use of 85% of the strength of the base metal (6005-T5) in the un-welded condition, provided that the assembly is artificially aged for 6 h at 182 °C (360 °F). Light poles fabricated from AA6063, welded in the T4 temper, and having a thickness either equal to or less than 9.5 mm, the specifications allow for use of 85% of strength of the base metal, *i.e.*, AA6063-T6, provided the welded assembly is artificially aged for 6 h at 182 °C (360 °F). It is important to note that the 85% percent "rule" is permissible and allowed when welding aluminum alloy 6005 and aluminum alloy 6063 using aluminum alloy 4043 as the filler material. The basis for these provisions within the Aluminum Design Manual (ADM) was the result of round robin tests carried out in the early 1960s and up to the late 1970s [9,10]. Most importantly, results of these studies were never published in the open literature, and some test records have been either misplaced or lost over the years. A careful review of the available test data from an earlier round robin test program, using statistical techniques inscribed within the 2010 Aluminum Design Manual (ADM), was considered to be both incomplete and inconclusive [9,10].

The focus of the current study was a determination of the mechanical properties of welded and artificially aged aluminum alloys 6061, 6063 and 6005A having thicknesses commensurate with what is currently being used in aluminum alloy light poles. Further, the study provided an opportunity to carefully examine the extrinsic influence of welding on intrinsic microstructural effects in an attempt to characterize the microstructure-mechanical property relationships. Much of the work examining the relationships can be found elsewhere [11,12]. This paper focuses on the mechanical properties of Post Weld Heat Treated (PWHT) aluminum alloys that are preferentially chosen for use in light poles. In recent years, the influence of post-weld heat treatment subsequent to hybrid welding of aluminum alloy 5754 was carefully studied and the test results and observations documented in the open literature [13].

2. SPECIMEN PREPARATION AND MECHANICAL TESTING

The parent materials chosen for use in this research study were the three aluminum alloys: (i) AA6063-T4, (ii) AA6061-T4, and (iii) AA 6005A-T1. All the three aluminum alloys were obtained in the as-extruded form. Blanks were then saw cut from the extruded sections. The tensile test specimens were precision machined from the as-extruded sections, the test specimen measured 6.4 mm (1/4 in) and 9.5 mm (3/8 in) in thickness, for both aluminum alloy 6063 and aluminum alloy 6061. Samples prepared for aluminum alloy 6005A were taken from extruded sections that measured 3.2 mm (1/8 in) in thickness. Different material thicknesses were selected based on their historical use in light pole applications, as well as provisions documented in the Aluminum Design Manual (ADM) for the purpose of enabling a scientific comparison of the tensile response.

The nominal compositions of the three alloys are given in Table 1 [14].

Table 1. Nominal composition (weight pct.) of the chosen 6XXX aluminum alloys [8].

Element	Al	Cr	Cu	Fe	Mg	Mn	Si	Ti	Zn
6063	Balance	0.1	0.1	0.35	0.9	0.1	0.6	0.1	Max 0.1
6005A	Balance	0.3	0.3	0.35	0.4	0.5	0.9	0.1	0.2
6061	Balance	0.04	0.15	0.7	1.2	0.15	0.8	0.15	Max 0.05

The filler metal used for this study was AA4043, as is common in the fabrication of light poles and for purpose of welding the 6XXX series alloys [15]. Fillet welds were used to form heat-affected zones across samples in an attempt to examine the influence of "localized" heating on the base material. The test samples, with short cover plates and a fillet weld, are shown in Figure 1. Gas metal arc welding (GMAW) was successfully used to deposit the weld metal using the technique of spray transfer, with the current varying from 120 to 220 A at a voltage of 22 V. The filler wire was 2.4 mm (3/32 in) in diameter with the shielding gas was 100 percent argon. Gas metal arc welding was chosen and used since it continues to be acceptable and preferentially chosen for use in a wide spectrum of industrial-related applications.

After the welds were placed, a Bridgeport vertical axis milling machine was used to remove the fillet and cover plate, leaving only the parent metal strip containing the heat affected zone (HAZ). Upon removal of the fillet welds and cover plates, each metal strip was transformed into a tensile specimen using a computer numerical control (CNC) machine (Model: HAAS) (Figure 2). A number of un-welded strips were also machined to provide tensile samples. A sizeable number of both the welded and parent metal tensile samples were chosen for the purpose of subsequent heat treatment.

Figure 1. Pictorial view of test sample with fillet welded lap joint.

The heat treatment process followed procedures outlined in ASTM B918-01 [16]. Essentially, a precipitation type heat treatment was used, and this is referred to as post-weld heat treatment (PWHT). For the three chosen aluminum alloys (6005A, 6061 and 6063), the specifics of the treatment involved an initial soak at a temperature of 182 °C (360 °F) for six hours, followed by cooling in ambient air (27 °C). In this research study, both the welded (PWHT) and unwelded (PHT) tensile samples were subject to artificial aging at 182 °C (360 °F) for six hours. The resultant tempers were expected to be near the T6 condition for aluminum alloy 6063 and aluminum alloy 6061. Test specimens fabricated from aluminum alloy 6005A were expected to be close to a T5 temper following artificial aging. The guaranteed minimum strengths for extrusions of AA6005A-T5, AA6061-T6 and AA 6063-T6 were 262 MPa, 262 MPa and 207 MPa, respectively. The three aluminum alloys chosen for this research study had strengths that exceeded the guaranteed minimum yield strength and ultimate tensile strength that is recommended for use in conventional structural design.

Figure 2. A finished tensile test specimen containing a fillet weld subsequent to machining on a Computer Numerical Control (CNC) machine.

A few tensile samples consisting of welded and parent metal AA6005A were re-solution heat treated and subsequently aged. The re-solution heat treatment followed guidelines given in Volume 4 of the ASM Handbook "Heat Treating" [17]. The re-solution treatment process consisted of an initial soaking of the test specimens at 529 °C (985 °F) for a full 60 min. This was followed by a rapid quench in a solution mixture of 60%/40% water/glycol. This treatment was expected to place both the coarse and intermediate-size constituent particles back in solution. If left unattended, the 6XXX-series alloys used in this research study would be expected to naturally age with time at an ambient temperature (27 °C). Following re-solution heat treatment, a selected number of specimens were subjected to artificial aging. The primary purpose of testing the re-solution heat treated and aged AA6005A samples was to compare the tensile properties with the 6005A counterpart.

Tensile samples of aluminum alloys 6061, 6063 and 6005A were tested in uniaxial tension in each of the following conditions:

1. As-Received (AR-6063, AR-6061 and AR-6005A).
2. As-Received and artificially aged (AR+PHT-6063, AR+PHT-6061, AR+PHT-6005A).
3. As-Welded (AW-6063, AW-6061, AW-6005A).
4. Welded and subsequently heat treated (PWHT-6061, PWHT-6063, PWHT-6005A).
5. As-Received and As-Welded subjected to re-solution heat treatment and aging (SHT+PHT-6005A).

Individual test specimens were placed in a universal test machine (Model: Warner-Swasey) and deformed in uniaxial tension up until failure by separation. An extensometer was fixed along the gage section of each sample to obtain a record of the axial strain during loading. Data from each test were recorded on a PC-based data acquisition system and subsequently used to develop the stress *versus* strain response. The engineering stress *versus* engineering strain curves were compared to provide an understanding of the response of the chosen test specimens when subjected to uniaxial deformation. Both yield strength and ultimate tensile strength values were obtained for each test sample, and the lower bounds statistically determined. The lower bound strengths were compared with the minimum guaranteed design values for the chosen aluminum alloy.

3. RESULTS AND DISCUSSION

3.1. Microstructure

Light optical micrographs were taken over a range of low magnifications, revealed the initial microstructure of AA6061 in the (a) as-received (AR); (b) as-welded (AW); (c) as-received plus precipitation heat treated (AR+PHT); and (d) post weld heat treated (PWHT) conditions. The as-received alloy revealed a random distribution of both coarse and intermediate size intermetallic particles (Figure 3a,b). These intermetallic particles result from the presence and availability of residual elements, such as iron and silicon [18,19]. As documented elsewhere, these particles were identified to be $Al_{12}Fe_3Si$, $Al_{15}(FeMn)_3Si$ and Al_5FeSi [18,19,20]. The iron-rich intermetallic particles range in size from 1 to 10 microns and are potential sites for the early initiation of microscopic damage during plastic deformation. The manganese-rich particles in the chosen aluminum alloys help in controlling both grain size and grain growth during solidification. Micrographs of the alloy in the as-welded (AW) condition revealed fine recrystallized grains in the region of the weld bead (Figure 4a). By comparison, the weld bead in the post weld heat treated condition is shown in Figure 4b. A noticeable difference in microstructure between the weld bead and base metal is evident along the interface between the two regions (Figure 5a,b). The microstructure of aluminum alloy 6061, in the as-received plus precipitation heat treated condition revealed a significant volume fraction of both coarse and intermediate size second-phase particles in the base metal (Figure 6a). These particles were randomly dispersed throughout the microstructure. The as-welded and post weld heat treated samples revealed very well defined grains that could be classified as being: (i) small in size and of varying shape (Figure 6b), and (ii) distributed randomly through the microstructure of the base metal. The microstructure at the interface of the base metal and weld bead is shown in Figure 12. Fine microscopic cracks are evident and can be attributed to melting of the low melting point constituents both at and along the grain boundaries.

3.2. Typical Stress-Strain Response

In Figure 7, Figure 8, Figure 9, Figure 10 and Figure 11 the typical stress *versus* strain behaviors for the aluminum alloys and test conditions employed used this study are shown. Each plot shows the stress *versus* strain variation for either the as-received (AR) and as-received and aged (AR+PHT) or the as-welded (AW) and the Post Weld Heat Treated (PWHT) counterpart. Figure 11, shows the stress *versus* strain response for the re-solution heat treated and aged 6005A (SHT+PHT or SHT+PWHT).

In general, artificial aging increased the strength of aluminum alloy 6063 for both the as-received (AR) and as-welded (AW) conditions. The observed increase in strength was far more pronounced for the aluminum alloy material that was 6.4 mm (1/4 in) thick. This is not unexpected, primarily because a thicker material necessitates the need for additional heat input during welding.

As in the case of AA6063, all tensile specimens of AA6061 responded positively to heat treatment, showing an observable gain in strength. The increase in strength was evident even for the 9.5 mm (3/32 in) thick specimens that were initially welded and subsequently aged. However, the ultimate tensile strength obtained for the thicker specimen (t = 9.5 mm) was found to be lower than the test specimens that measured 6.4 mm (1/4 in) in thickness.

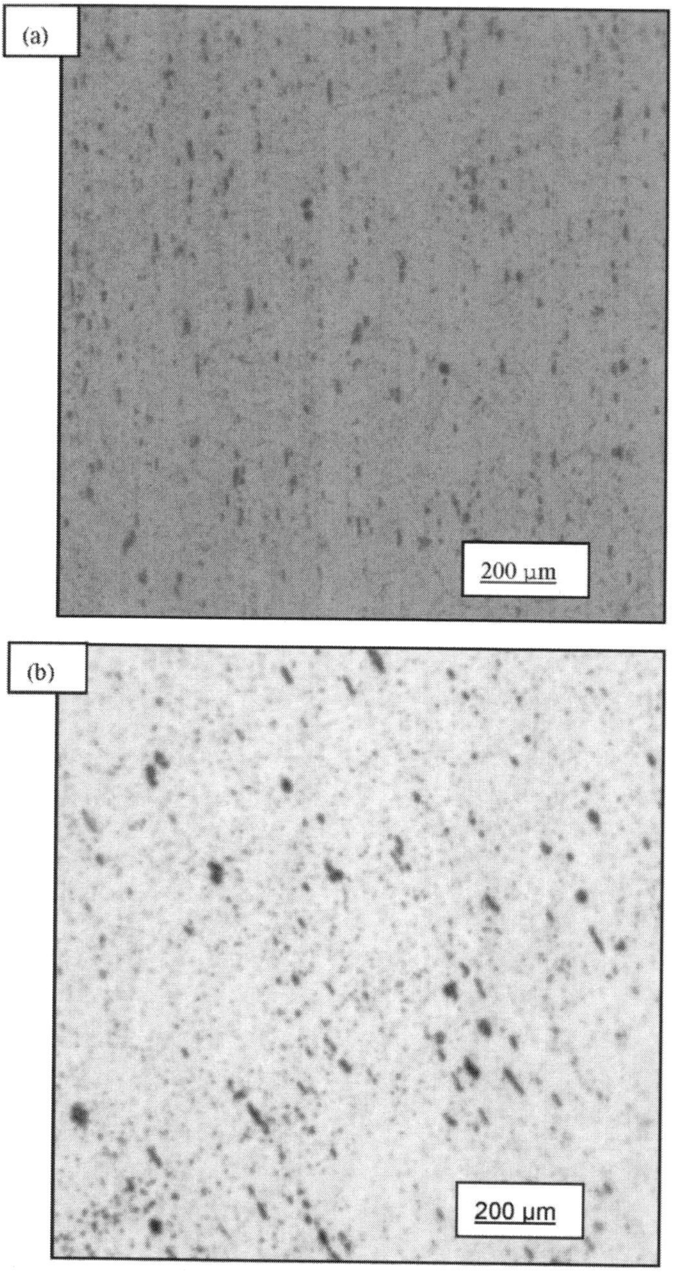

Figure 3. Light optical micrographs of aluminum alloy 6061-T4 showing microstructure the following: (**a**) Coarse and intermediate second phase particles in the base metal of the as-received or as-provided metal; (**b**) distribution of intermetallic particles in the heat-treated sample.

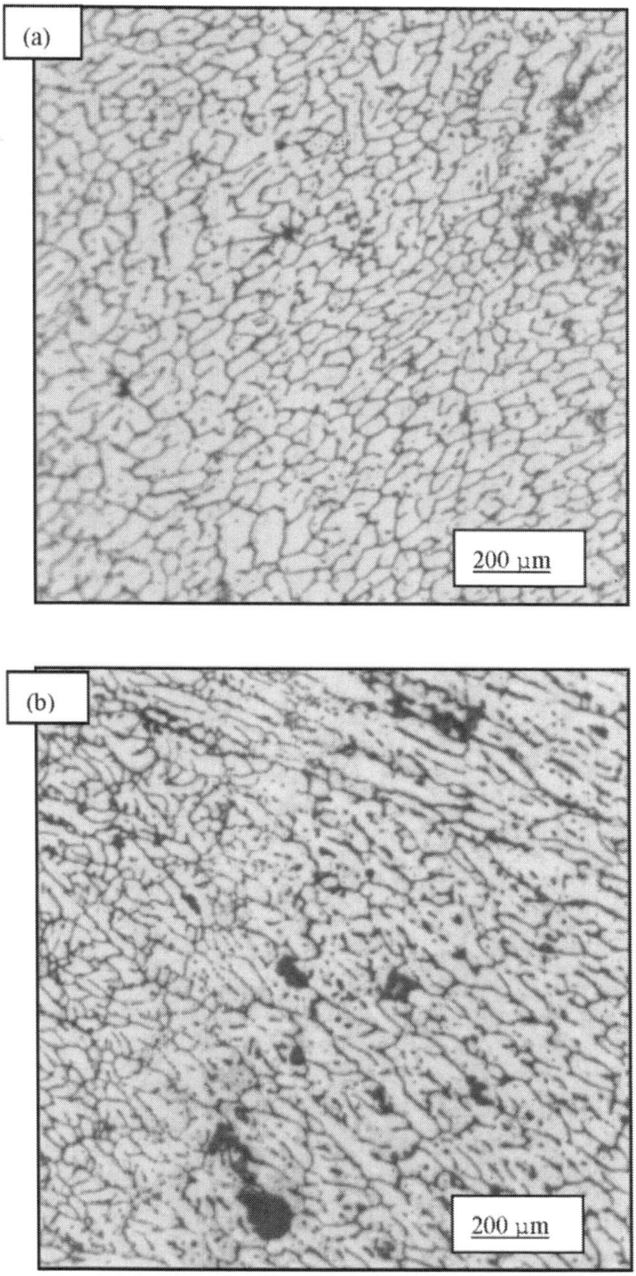

Figure 4. Optical micrograph of the weld pool showing fine grains of varying size and shape. (**a**) Grain size and morphology in the weld pool in the as-welded condition; (**b**) weld pool in the post weld heat-treated condition.

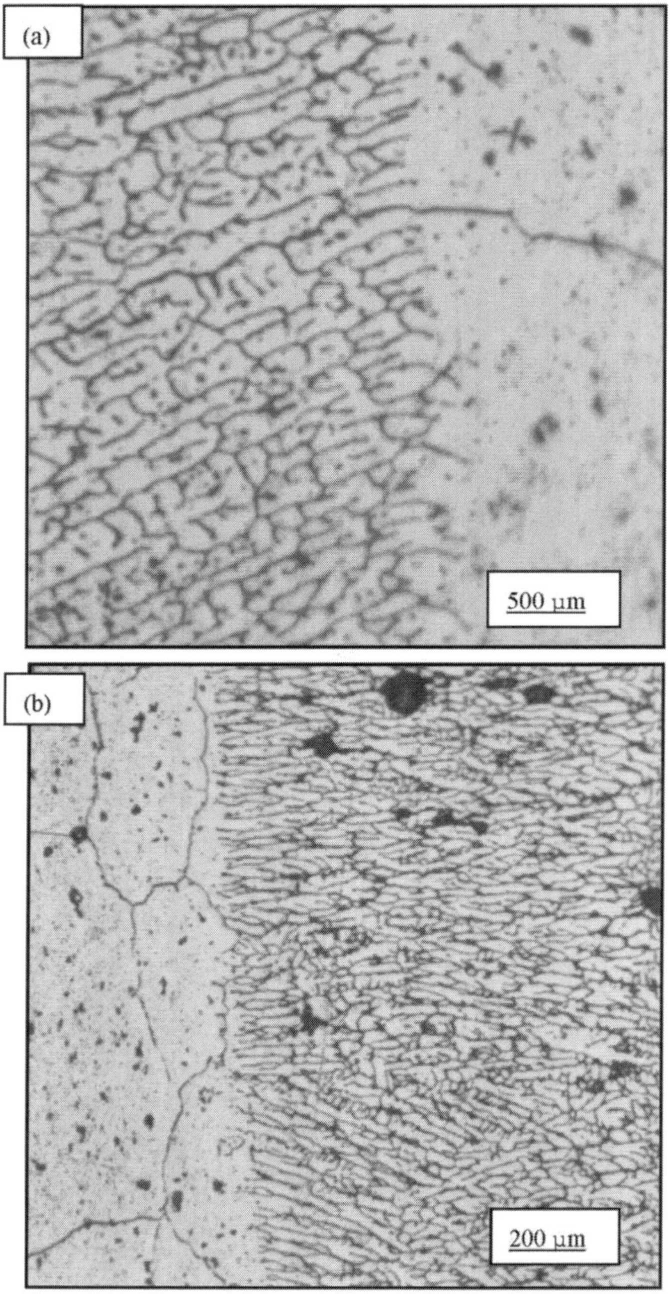

Figure 5. Optical micrographs showing the following: (**a**) Microstructure at the weld-base metal interface of the as-welded Aluminum alloy 6061-T4; (**b**) microstructure of the weld-base metal interface in the post weld heat treated aluminum alloy 6061.

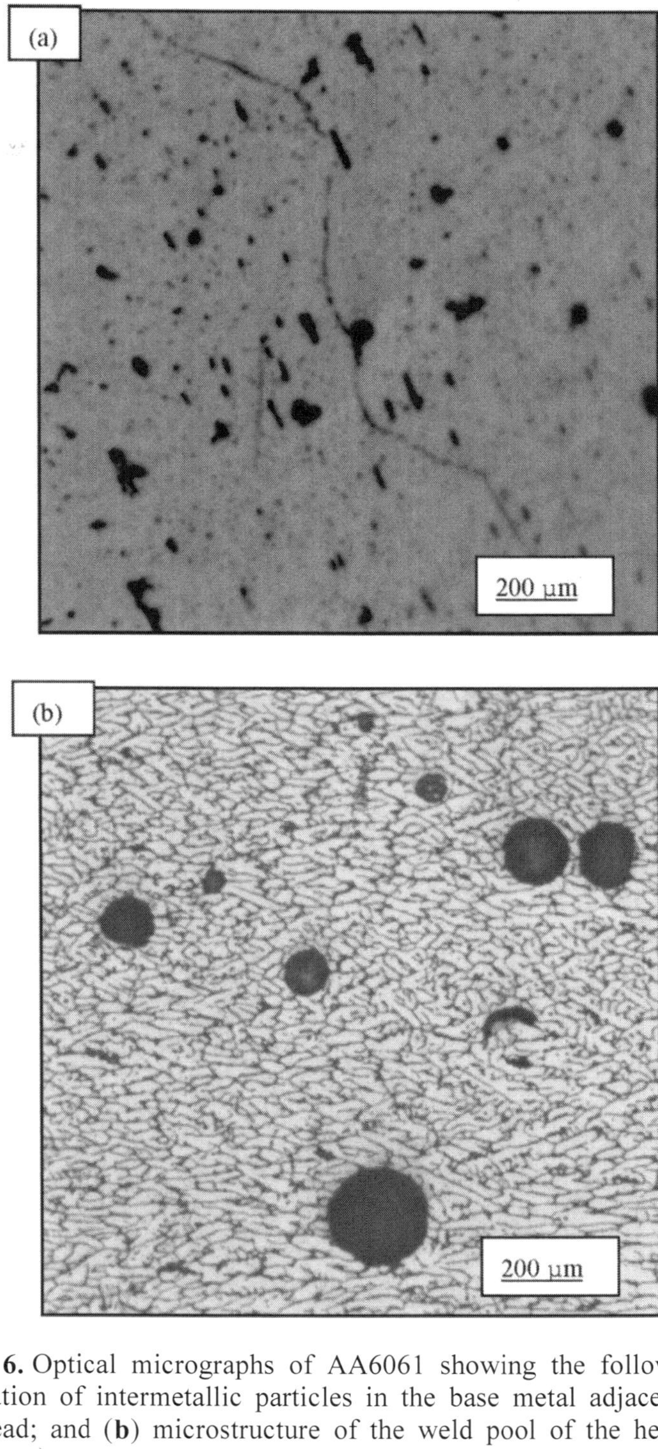

Figure 6. Optical micrographs of AA6061 showing the following: (**a**) Distribution of intermetallic particles in the base metal adjacent to the weld bead; and (**b**) microstructure of the weld pool of the heat-treated alloy.

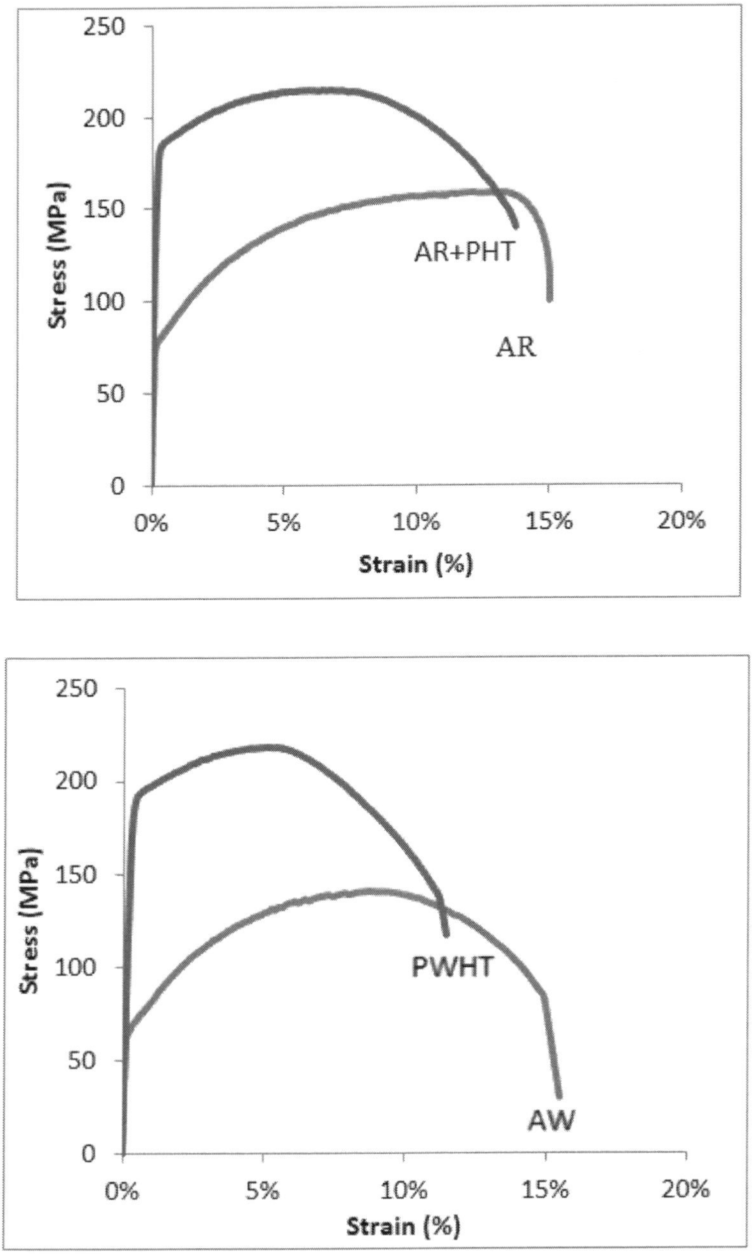

Figure 7. Stress *versus* strain response of 6.4 mm (1/4 in) thick AA6063 in the as-received (AR), as-received plus precipitation heat treated (AR+PHT), as-welded (AW) and post weld heat treated (PWHT) conditions.

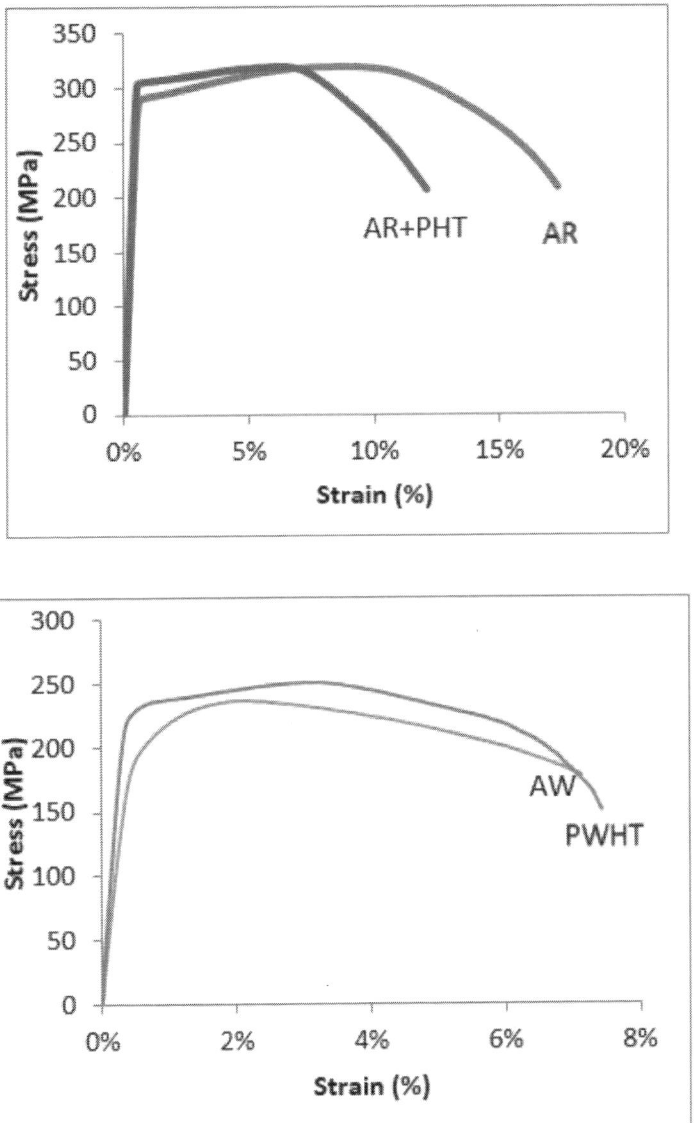

Figure 8. Stress *versus* Strain response of 9.5 mm (3/8 in) thick AA6063 in the as received (AR), as-received plus precipitation heat treated (AR+PHT), as-welded (AW) and post weld heat treated (PWHT) conditions.

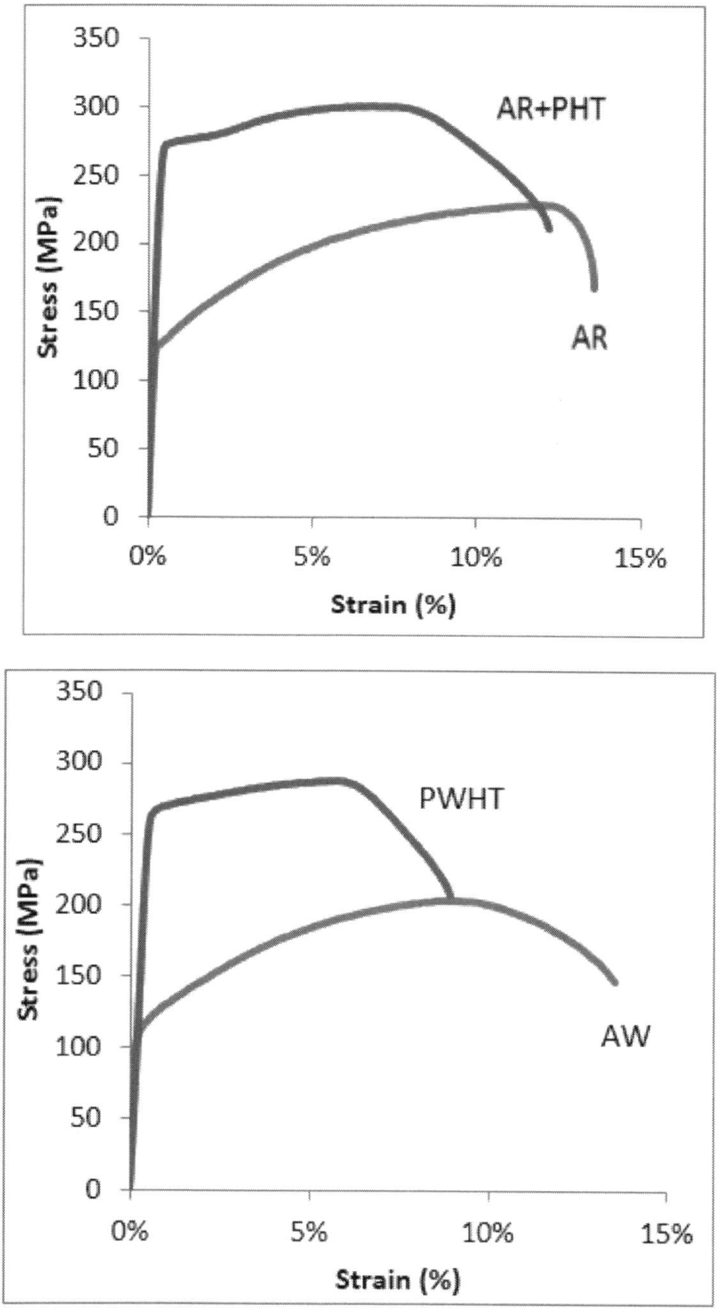

Figure 9. Stress *versus* strain response of 6.4 mm (1/4 in) thick extrusion of AA6061 in the as-received (AR), as-received plus precipitation heat treated (AR+PHT), as-welded (AW), and post weld heat treated conditions.

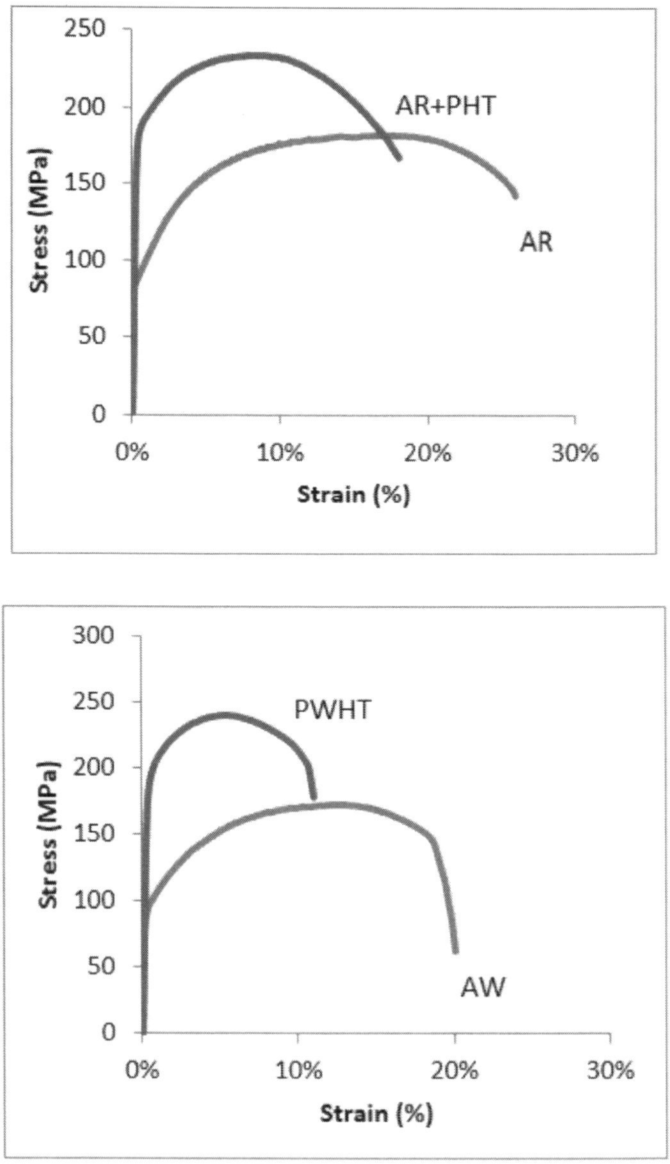

Figure 10. Stress-strain response of 9.5 mm (3/8 in) thick extrusion of AA6061 in the as-received (AR), as-received plus precipitation heat treated (AR+PHT), as-welded (AW) and post weld heat treated (PWHT) conditions.

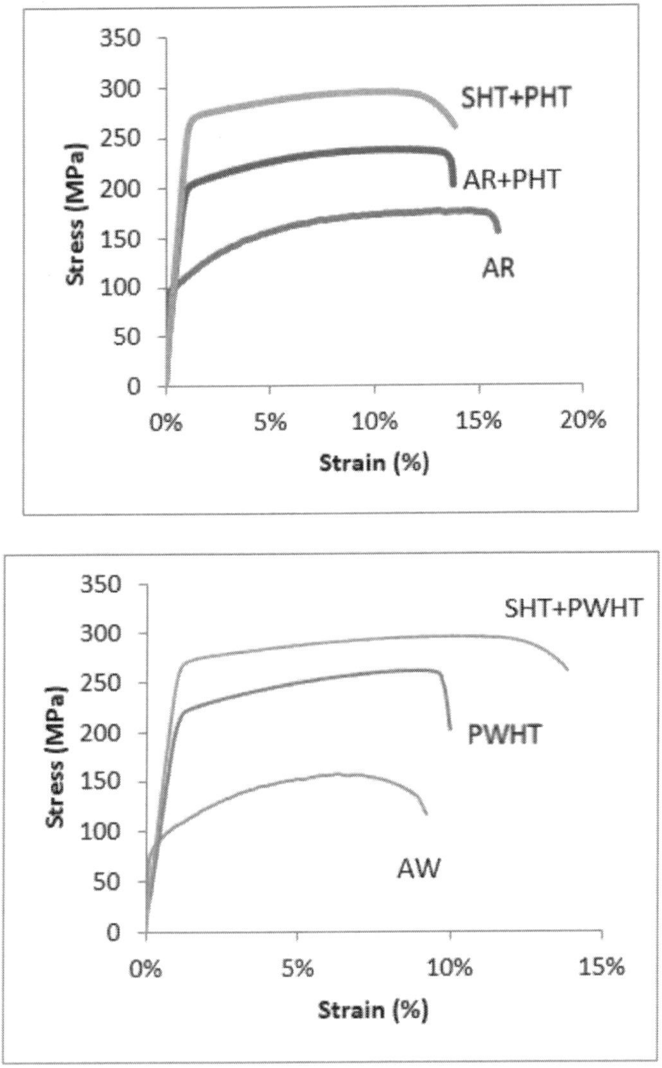

Figure 11. Stress *versus* strain response of 3.2 mm (1/8 in) AA6005A in the as-received (AR), as-received plus precipitation heat treated (AR+PHT), solution heat trearted plus precipitation heat treated (SHT+PHT), as-welded (AW), post weld heat treared (PWHT) and solution heat treated plus post weld heat treated (SHT+PWHT) conditions.

Both as-received (AR) and as-welded (AW) specimens of AA6005A, having a thickness of 3.2 mm (1/8 in), responded favorably to heat treatment, with a significant increase in both yield strength and tensile strength. Test specimens of aluminum alloy 6005A, in both the as-received (AR) and as-welded (AW) conditions, that were solution heat treated and subsequently aged showed the largest gain in strength when deformed in uniaxial tension.

3.3. Analysis of the Results

Welding did have an influence on both microstructure and mechanical properties of the chosen 6XXX series aluminum alloys. A similar influence of welding, *i.e.*, hybrid welding, was observed to have a noticeable influence on weld quality, microstructure and mechanical properties of aluminum alloy 5754 [13] and an experimental Al–Mg alloy [21]. The influence of welding differs depending on the following: (a) the alloy chosen; (b) the welding process used; (c) the parameters employed; and (d) overall quality of the weld. The type of joint and thickness of the starting material does have an influence on heat input, microstructure and resultant strength. Not surprisingly, this study of 6063, 6061 and 6005A aluminum alloys revealed that selective artificial aging or heat treatment increased the mechanical strength of the alloys. The observed increase in strength for the PWHT samples can be attributed to the existence of diffusion-assisted mechanisms that favor an initial increase in Guinier Preston (GP) zones coupled with a hinderance caused to the movement of dislocations as a consequence of the formation and presence of matrix strengthening precipitates. The precipitates in the PWHT alloy are finer and more uniformily distributed in the aluminum alloy metal matrix. This favors an increase in dislocation density, which contributes to the observed improvement in both yield strength and tensile strength.

When subjected to “localized” heat input as a direct consequence of welding, the chosen aluminum alloys experienced a decrease in strength when compared to strength of the as-received condition. The decrease in strength can be essentially attributed to changes in intrinsic microstructural features of the starting material as a consequence of the heat input during welding. A majority of the as-welded (AW) samples broke in an area adjacent to the weld; normally on the side of the weld to which more heat was applied. The heat-affected zone (HAZ) was observed to have lower strength when compared to the base metal.

A statistical analysis, using the guidelines established in the 2010 edition of the Aluminum Design Manual, coupled with the published guaranteed minimum strengths for AA6061, AA6063 and AA6005A, was used to determine reasonable design minimum strength for the samples that were subject to post weld heat treatment. The minimum tensile strength and yield strength for the PWHT tensile samples was established using the following equation:

$$\sigma_{min} = \sigma_{avg} - kS_{\sigma} \quad (1)$$

where

- σ_{min} = calculated minimum stress for the PWHT specimens
- σ_{avg} = average tensile or yield strength for a given alloy in the PWHT condition
- k = statistical coefficient based on the number of tests, n
- S_{σ} = standard deviation of the test results for the particular alloy

Results of the analysis of the strength of the PWHT specimens are summarized in Table 2. Detailed in Table 2 are: (a) the alloys studied; (b) thicknesses; (c) average yield strength and average ultimate tensile strength; (d) the number of tests in the data set; (e) the standard deviation; (f) the calculated minimum yield strength and ultimate tensile strength; (g) the ratio of the calculated minimum yield strength to the guaranteed minimum yield strength for the base alloy; as well as (h) ratio of the calculated minimum ultimate strength to the guaranteed minimum strength for the base alloy. Ratio of the calculated minimum yield strength to guaranteed minimum yield strength varied from 0.5 to a high of 0.99. Ratio of the calculated minimum ultimate strength to guaranteed strength varied from 0.66 to a high of 0.98. The lower values (0.5 and 0.68) correspond to 6061 having a thickness of 9.5 mm (3/8 in) that was post weld heat treated and tested. In order to obtain test specimens of aluminum alloy 6061 that were 9.5 mm (3/8 in) thick, blanks were removed from an extrusion that had a initial thickness of 12.7 mm (1/2 in), and subsequently milled to the required thickness of 9.5 mm (3/8 in). While this may have had some influence on the test results, Table 2 shows that for both AA6063 and AA6061, the data for the 9.5 mm (3/8 in) thick specimens had noticerably larger standard deviations for both yield strength and ultimate tensile strength when compared to the specimens that measured 6.4 mm (1/4 in) in thickness. This is attributed to the increased heat input used during the welding of the thicker materials. Presented and discussed in detail elsewhere [18], an examination of the microstructure of the area immediately around the fillet revealed a few instances of visible secondary melting (Figure 12).

Table 2. Data analysis summary—post weld heat treated (PWHT) yield and ultimate strengths.

Alloy	Thick (mm)	Average Yield (MPa)	Average Ultimate (MPa)	Number of Tests	Standard Deviation Yield/Ultimate	Min. Yield (MPa)	Min. Ultimate (MPa)	Min. Yield/ADM	Min. Ult./ADM
6063	6.4	190	219	10	0.92/0.95	165	193	0.96	0.93
6063	9.5	204	234	21	2.44/1.57	150	199	0.87	0.96
6005A	3.2	227	272	12	1.0/1.05	201	245	0.83	0.93
6005A	6.4	217	250	12	0.44/0.82	205	229	0.85	0.87
6061	6.4	261	281	15	0.93/0.98	239	258	0.99	0.98
6061	9.5	241	241	24	4.1/3.1	122	173	0.5	0.66

Note: Minimum Yield/ADM = calculated minimum yield strength from the testing program divided by the guaranteed minimum yield strength for the alloy. Minimum Ultimate/ADM = calculated minimum ultimate strength from the testing program divided by the guaranteed minimum ultimate strength for the alloy.

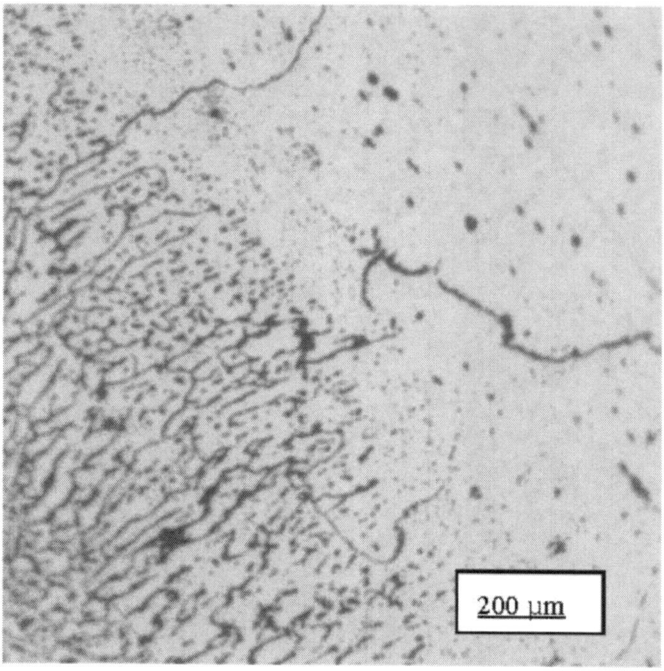

Figure 12. Optical micrograph showing region experiencing secondary melting in the weld-base metal fusion line in aluminum alloy 6061.

For the remaining alloys and thicknesses chosen, the provisions within the 2010 Aluminum Design Manual allowing the use of 85%. of the parent metal strength for PWHT light poles are confirmed. In addition, using 85%. of the base metal strength for extruded aluminum alloy AA6061, for thicknesses up to 6.4 mm (1/4 in) and welded in the T4 temper using AA4043 as filler followed by post weld heat treatment (PWHT), also conforms with the 85% rule. In summary the alloys and thicknesses include the following:

1. 6063-T4, PWHT up to 9.5 mm (3/8 in) thick and welded using AA4043.
2. 6005A-T1, PWHT up to 6.4 mm (1/4 in)thick and welded using AA4043.
3. 6061-T4, PWHT up to 6.4 mm (1/4 in) thick, and welded using AA4043.

4. CONCLUSIONS

This study examined the influence of post weld heat treatment (PWHT) on strength of three aluminum alloys that are commonly chosen for use in welded light poles. Findings of the study are as follows:

1. Heat treating (aging) the as-received (AR) material increased both the yield strength and ultimate tensile strength of all the alloys.
2. Post weld heat treating increased both the yield strength and ultimate tensile strength of the three alloys studied.
3. Re-solution heat treating the as-received material increased the yield strength and tensile strength of aluminum alloy 6005A.
4. Re-solution heat treating subsequent to welding, followed by post weld heat treatment was observed to increase the tensile strength and yield strength of aluminum alloy 6005A.
5. With the exception of aluminum alloy 6061 having a thickness of 9.5 mm (3/8 in), design provisions permitting use of 85% of the parent metal strengths (in T6 temper) for post weld heat treated (PWHT) light poles are confirmed. The alloys and thicknesses include: (i) 6063-T4 PWHT up to 9.5 mm thick; (ii) 6005A-T1 PWHT up to 6.4 mm thick; and (iii) 6061-T4 PWHT up to 6.4 mm thick.

ACKNOWLEDGMENTS

This research was made possible through funds provided by the University of Akron for the graduate student and with material and technical support from Hapco Inc (Abingdon, VA, USA).

AUTHOR CONTRIBUTIONS

Eric Hilty performed the specimen fabrication and conducted the tests. Menzemer conceived and developed the program as well as advised Hilty. He also performed the final data analysis and assisted with manuscript preparation. Srivatsan conducted examination of the material microstructures, examined the fracture surfaces and assisted with the manuscript preparation. Morrison and Minor provided the material and guidance to the study and assisted with manuscript preparation.

CONFLICTS OF INTEREST

The authors declare no conflict of interest.

REFERENCES

1. Sharp, M.L. *Behavior and Design of Aluminum Structures*; McGraw-Hill: New York, NY, USA, 1992.

2. Kissell, R.; Ferry, R. *Aluminum Structures—A Guide to Their Specifications and Design*; John Wiley and Sons: London, UK; Washington, DC, USA, 1995.

3. Elangovan, K.; Balasubramanian, V. Influence of post weld heat treatment on tensile properties of friction stir welded AA6061 aluminum alloy joints. *Mater. Charact.* **2008**, *59*, 1168–1177.

4. Demir, H.; Gunduiz, S. The effects of aging on the machinability of 6061 aluminum alloy. *J. Mater. Des.* **2009**, *30*, 1480–1483.

5. Xu, W.; Gittos, M.F. *Materials and Structural Behavior of MIG Butt Welded in 6005-T6 Aluminum Alloy Extrusions under Quasi Static and Impact Loading*; TWI Report Number 14054/1/04; The Welding Institute: Cambridge, UK, 2004.

6. Groover, M.P. *Fundamentals of Modern Manufacturing, Materials, Processes and Systems*, 3rd ed.; John Wiley and Sons: Hoboken, NJ, USA, 2007.

7. Hatch, J.E. *Aluminum: Properties and Physical Metallurgy*, 10th ed.; American Society for Metals: Metals Park, OH, USA, 1983.

8. The Aluminum Association. *Aluminum Design Manual*; Aluminum Association of America: Washington, DC, USA, 2010.

9. Maedler, J.R. *NEMA to Gordon Allison*; The Aluminum Association: Washington, DC, USA; 12; February; 1971.

10. Wagoner, N. *Reynolds Metals Company to R*; Hartmann, Hapco: Abingdon, VA, USA; 5; August; 1963.

11. Hilty, E.; Menzemer, C.; Srivatsan, T. Influence of Welding and Heat Treatment on Microstructural Development and Properties of Aluminum Alloy 6005. In Proceedings of the 22nd PFAM Conference, Singapore, 18–20 December 2013.

12. Hilty, E.; Menzemer, C.; Manigandan, K.; Srivatsan, T. Influence of welding and heat treatment on microstructure, properties and fracture behavior of a wrought aluminum alloy. *Emerg. Mater. Res.* **2014**, *3*, 230–242.

13. Leo, P.; D'Ostuni, S.; Casalino, G. Hybrid welding of AA5754 annealed alloy: Role of post weld heat treatment on microstructure and mechanical properties. *Mater. Des.* **2016**, *90*, 777–786.

14. The Aluminum Association. *2009 Aluminum Standards and Data*; Aluminum Association of America: Washington, DC, USA, 2009.

15. Hobart Filler Metals. *Guide for Aluminum Welding*; Illinois Tool Works: Chicago, IL, USA, 2013.

16. ASTM B918-01. *Standard Practice for Heat Treatment of Wrought Aluminum Alloys*; ASTM International: West Conshohocken, PA, USA, 2001.

17. *ASM Materials Handbook*; ASM International: Materials Park, OH, USA, 1991; Volumes 4 and 6.

18. Warmuzek, M.; Mrowkaand, G.; Sieniawski, J. Influence of Heat Treatment on Precipitation of Intermetallic Phases in Commercial AlMnFeSi Alloy. *J. Mater. Process. Technol.* **2004**, *157–158*, 624–632.

19. Gallais, C.; Simar, A.; Fabreque, D.; Denquin, A.; Lapasset, G.; de Meester, B.; Brechet, Y.; Pardoen, T. Multiscale Analysis of the Strength and Ductility of AA 6056 Aluminum Friction Stir Welds. *Metall. Mater. Trans.* **2007**, *38*, 964–981.

20. Starke, E.A., Jr. *Fatigue and Microstructure*; Meshii, M., Ed.; ASM International: Materials Park, OH, USA, 1979.

21. Leo, P.; Renna, G.; Casalino, G.; Olabi, A.G. Effect of power distribution on weld quality during hybrid laser welding of an Al–Mg alloy. *Opt. Laser Technol.* **2015**, *73*, 118–126.

CHAPTER 5

Effects of Cryogenic Forging and Anodization on the Mechanical Properties and Corrosion Resistance of AA6066–T6 Aluminum Alloys

Teng-Shih Shih [1,*], Hwa-Sheng Yong [1,2] and Wen-Nong Hsu [1,2]

[1] Department of Mechanical Engineering, National Central University, Jhongli District, Taoyuan City 32001, Taiwan

[2] Graduate Student, National Central University, Jhongli District, Taoyuan City 32001, Taiwan

ABSTRACT

In this study, AA6066 alloy samples were cryogenically forged after annealing and then subjected to solution and aging treatments. Compared with conventional 6066-T6 alloy samples, the cryoforged samples exhibited a 34% increase in elongation but sacrificed about 8%–12% in ultimate tensile strength (UTS) and yield stress (YS). Such difference was affected by the constituent phases that changed in the samples' matrix. Anodization and sealing did minor effect on tensile strength of the 6066-T6 samples with/without cryoforging but it decreased samples' elongation about 8%–10%. The anodized/sealed anodic aluminum oxide (AAO) film enhanced the corrosion resistance of the cryoforged samples.

KEYWORDS

cryoforging; anodization; tensile properties; corrosion resistance

1. INTRODUCTION

Al–*x*Mg–*y*Si alloys (6xxx series Al alloys) are commonly used as extruded shapes and forged for making bicycle parts. Their characteristics include ample formability, machinability, weldability, and corrosion resistance, as well as good strength and elongation after heat treatment. These alloys are also readily available on the market.

Aluminum possesses a high stacking fault energy and readily undergoes dynamic recovery during deformation. Plastic deformation at low temperatures, such as cryorolling, is beneficial for refining grains in an aluminum alloy matrix [1,2]. Chen studied equal-channel deformation of an Al–Mg alloy at cryogenic temperatures and found that high-density dislocations distorted grains to a refining grain size [3]. Lee *et al.* also found that cryorolling 5083 alloy could obtain 200 nm fine grains to increase the ultimate tensile strength (UTS) from 315 to 522 MPa [4]. For an Al–Mg–Si alloy, cryorolling with a 90% reduction could cause heavy plastic deformation to produce nanosized extra-fine grains [5,6,7].

Aluminum alloys that become deformed at cryogenic temperatures could suppress the dynamic recovery that occurs during plastic deformation, and could acquire fine grains featuring high-angle grain boundaries [8,9]. The interaction of high-density dislocations enhanced the precipitation capability for producing a high density of nanosized precipitates [10]. Yin *et al.* found nanograins that were less than 500 nm in size in 7075 alloy samples subjected to compression forging at cryogenic temperatures [11].

Krishina *et al.* [8,12] produced ultrafine-grained Al–4Zn–2Mg alloys by cryorolling and indicated that the driving force for precipitation could be enhanced by differential scanning calorimetry. As a result, the precipitates of the η phase became finer compared with conventional aging treatment. Sarma [13] found that cryorolling significantly changed aging behavior, leading to a reduction in the aging temperature from 190 to 125 °C and in aging time from 24 to 8 h for treating 2219 alloy (Al–Cu alloy).

Jayaganthan [14] used X-ray analyses to study the aging behavior of 6061 alloy subjected to solution treatment then cryorolling. They found that increasing true strain in cryorolling tended to enhance the dissolution of alloys in the alloy matrix and promoted the driving force for precipitation. The UTS was improved from 300 to 365 MPa and elongation was raised from 11% to 13%. Cryorolling apparently reduced the size of intermetallic compounds contained in the matrix of 7075–T73 alloys. As a result, the corrosion resistance of anodized and sealed 7075–T73 alloy could be significantly enhanced [15].

Anodization and sealing improves the corrosion resistance of Al alloys by forming amorphous alumina and hydrate alumina in the anodized film. This process has been widely used in industry. Forging is a common process used for making bicycle and automobile parts. Determining the influence of cryoforging on the corrosion resistance of Al alloys with or without anodization should provide more values for designing and using Al alloys.

Copper was added as an alloying element in a 6xxx series alloy to enhance mechanical strength by precipitation hardening after heat treatment. For example, 6066 Al alloy contains some high-strength Cu (0.8–1.4 Mg, 0.9–1.8 Si, 0.6–1.1 Mn, and 0.7–1.2 Cu) to obtain a UTS of 393 MPa and a yield stress (YS) of 359 MPa after a T6 treatment. This study introduced cryoforging to further improve the toughness of 6066–T6 alloys and their corrosion resistance. The effects of anodization and sealing on the tensile properties and corrosion resistance of AA6066–T6 with and without cryoforging were also evaluated.

2. EXPERIMENTAL PROCEDURES

2.1. Materials

As-extruded 6066 alloy bars, Ø42 × 100 mm in size, were supplied by Tzan Wei Aluminum Co., Ltd. (Tainan, Taiwan). The chemical compositions of the alloys (in wt. %) were 1.38 Si, 0.15 Fe, 1.18 Cu, 1.00 Mn, 1.09 Mg, 0.16 Cr, 0.04 Zn, and 0.02 Ti.

After annealing at 688 K for 120 min, the samples were divided into two groups. The first group was subjected to a solution treatment (803 K for 120 min) and artificial aging (450 K for 480 min); these samples were coded as T6 samples. The second group was subjected first to cryogenic forging, achieving a 40% reduced thickness (from 28 to 16.8 mm), then immersed in liquid nitrogen again, rotated by 90°, and subjected to a second round of cryogenic forging. A 500-ton hydraulic press equipped with one open die set was used to conduct compression forging. After forging, the second group of samples was subjected to the solution to get CFT4 sample and followed by artificial aging treatments: 450 K for 480 min for CFT6a samples and 540 min for CFT6b samples.

2.2. Tensile and Fatigue Tests

The specimens used for testing tensile and rotating bending fatigue strengths were machined from heat-treated samples according to the ASTM B557 [16] (gage diameter: 6 mm) and JIS Z2274 [17] (gage diameter: 8 mm) specifications as shown inFigure 1a,b. The machined test bars were polished by a series of abrasive papers (2000 grit) and an alumina slurry to achieve a surface roughness of less than 0.1 μm (*Ra*).

(a)

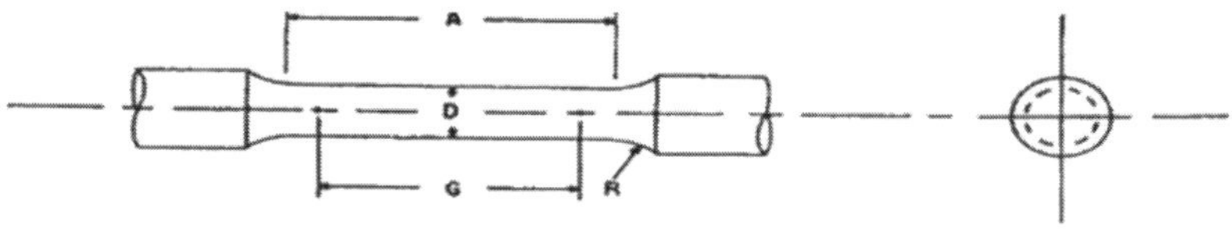

Nominal Diameter	Dimensions,mm
G-gage length	30±0.06
D-Diameter	6.00±0.10
R-Radius of fillet	6
A-Length of reduced section	36

(b)

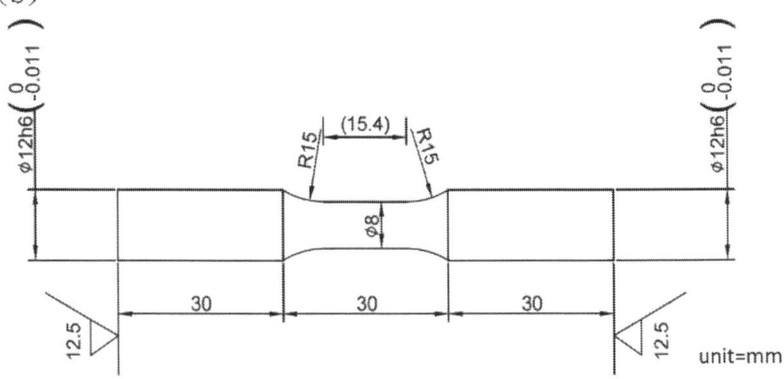

Figure 1. The dimensions of specimens used for (**a**) tensile test (gage diameter: 6 mm); and (**b**) rotating bending test (gage diameter: 8 mm).

2.3. X-ray Tests

X-ray diffraction (XRD) measurements were performed by using a NANO-Viewer Advance (Rigaku, Tokyo, Japan) equipped with a Cu target to identify the precipitates (or second-phase particles) formed in the matrix of different samples, including the T6, CFT4, and CFT6a samples.

The cryoforged samples were solution-treated to get a CFT4 sample and followed by aging treatment to acquire a CFT6a sample. A power of 30 KV and current of 10 mA were used in this study. The sample sizes were 10 × 10 × 1 mm.

2.4. Anodization Process

Before anodization, all samples were polished to a surface roughness of approximately $Ra \leq 0.1$ μm and then dipped into methanol and ultrasonically vibrated. The specimens were initially degreased by immersion in an alkaline solution (5 mass % NaOH) at 60 °C for 30 s and were then rinsed with water for 1–2 min. For the pickling process, specimens were submerged in an aqueous solution of HNO_3 (30 vol. %) for 90 s at room temperature and then rinsed with water for 1–2 min. The anodization was conducted at 15 $mA \cdot cm^{-2}$ at 15 °C for 900 s in a 15 mass % sulfuric acid solution. The anodized samples were sealed in hot water at 95 °C for 1200 s [18]. After anodization, scanning electron microscopy was conducted and determined that the anodic film was 12–14 μm thick. The anodized samples were sealed in hot water at 368 K for 20 min.

3. RESULTS AND DISCUSSION

3.1. Microstructure Observation and Tensile Properties

Table 1 shows the measured mechanical properties of different samples. The CFT6b samples, which aged for 60 min longer than did the CFT6a samples, had increased strength but reduced elongation. The CFT6a samples exhibited a similar strength but superior elongation to the CFT6b samples, and they were adopted in this study for evaluation of their corrosion resistance and fatigue strength. In addition, the CFT6a samples obtained a matrix with a uniform hardness (HV minimum deviation: 1.7).

Table 1. Mechanical properties of T6, CFT6a and CFT6b samples.

Sample	Tensile Strengths			Vickers Hardness (HV)
	UTS (MPa)	YS (MPa)	Elongation (%)	
T6	460 (0.5)	438 (1.4)	10.6 (1.7)	133.6 (2.9)
CFT6a	431 (6.6)	394 (9.2)	14.2 (0.2)	136.6 (1.7)
CFT6b	435 (5.4)	404 (10.1)	13.4 (0.9)	137.1 (2.9)

Note: the deviations are listed in parentheses.

Figure 2a,b show the second-phase (SP) and/or intermetallic compounds (IMC) located at the cross-sections in the transverse direction of the T6 and CFT6a samples. Different sizes of particles were counted and are listed in Table 2. The T6 samples exhibited more coarse SP/IMC particles than the CFT6a samples did as revealed by arrows in Figure 2. During the cryogenic forge, a shear stress was generated to act on the samples' matrix, breaking down the particles. As a result, the particles became finer and dissolution was enhanced during the solution treatment. The total SP/IMC particle count decreased from 864 to 734 counts/mm^2.

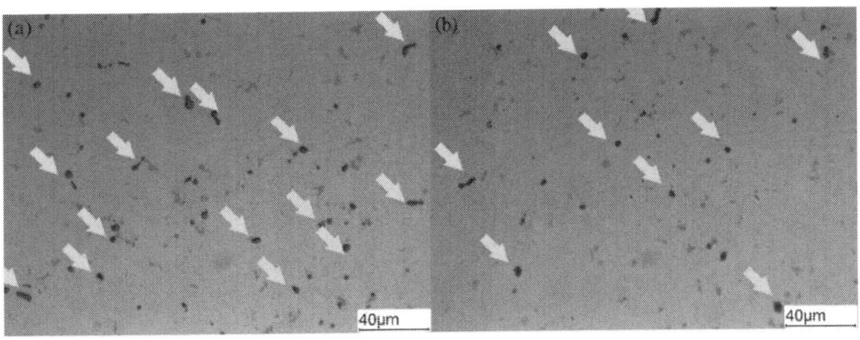

Figure 2. Optical micrographs show second-phase particles located in the matrix of (**a**) T6; (**b**) CFT6a sample; non-etched cross-section in transverse direction.

Table 2. The SP and/or IMC particles measured from T6 and CFT6a samples; maximum particle size and count population were included.

Sample	Second Phase (Coarse Precipitates) Counts/mm^2				Max. Diameter of Particle, μm
	1–10	11–20	>20	Total Count	
T6	838	25	0	864	17
CFT6a	719	15	0	734	17

Electron backscattered diffraction (SUPRA ULTRA 55 field emission scanning electron microscope, ZEISS, Jena, Germany) was performed to measure the misorientation angles of grain boundaries in the longitudinal direction of the tensile test bar samples, as illustrated in Figure 3a,b for the CFT6a and T6 samples, respectively. The CFT6a sample had a high fraction of high-angle grain boundaries (HAGBs). The matrix of the T6 sample had grains featuring mainly low-angle grain boundaries, as shown in Figure 3b. The cyclic loaded sample after being subjected to 250 MPa and fractured at 2.56 × 10^5 life cycles is shown in Figure 3c, revealing further increasing HAGBs compared with those in Figure 3a. This increase could have been affected by dynamic recrystallization during cyclic loading.

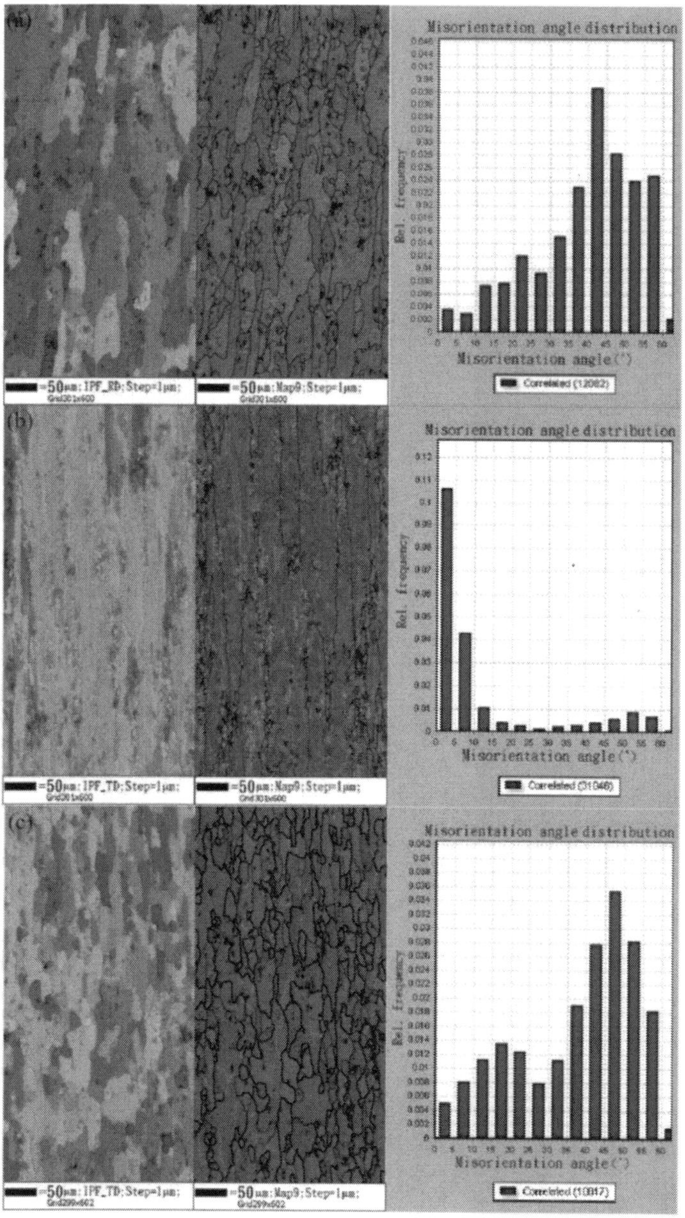

Figure 3. Inverse pole figure maps obtained from EBSD data and measured misorientation angle of grain boundaries from (**a**) CFT6a tensile tested sample; (**b**) T6 tensile tested sample; and (**c**) CFT6a sample after subjected to 250 MPa and fractured at 2.56 × 10^5 life cycles.

The T6 sample comprised α-Al, *Q*-(AlCuMgSi), Al(Mn,Fe)Si, and some Mg_2Si phases, as shown in Figure 4. After solution treatment, the cryoforged sample was tested and indicated that SP/IMC particles were highly dissolved in its matrix, as confirmed by the CFT4 sample in Figure 4. After artificial aging, the intensity of the *Q* phase was notably decreased, presenting complex phases of Mg_2Si, Q-(AlCuMgSi), and Al(Mn,Fe)Si, as well as some Cu_9Al_4 and CuMgSi phases; see CFT6a sample. Kim *et al.* annealed copper wire and an aluminum pad at 150 to 300 °C and found a Cu_9Al_4 phase with an X-ray spectra peak at 43.9° [19].

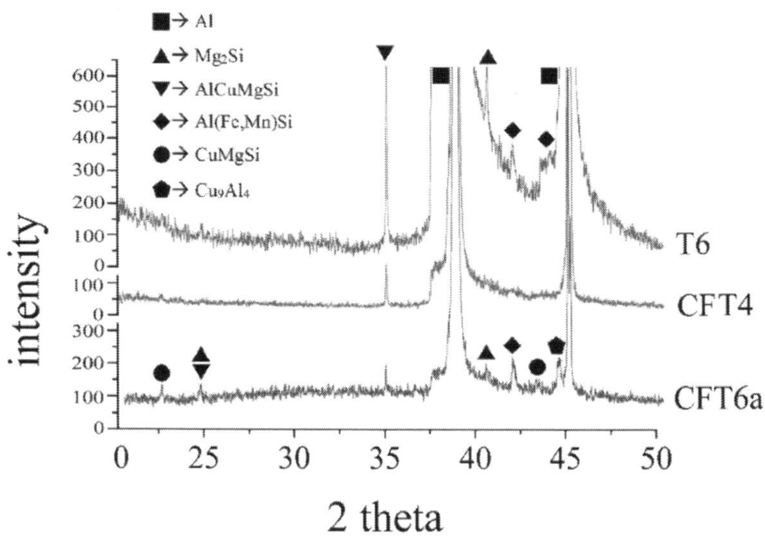

Figure 4. XRD patterns of 6066 alloy samples; including solution and aging T6 sample; cryoforged samples after solution treatment CFT4 and aging CFT6a, respectively.

The diffusivity of alloying elements in the aluminum matrix is in the order of $D_{Cu/Al}$ (4.44 × 10^{-5} m^2/s), $D_{Mg/Al}$ (1.49 × $10^{-5} m^2/s$), and $D_{Si/Al}$ (1.38 × 10^{-5} m^2/s) in face-centered cubic Al [20]. Cu atoms tend to segregate around the metastable phase in the Al–Mg–Si alloy, move to grain boundaries during the solution treatment, and diffuse to Mg–Si nanoparticles located at or near grain boundaries to finally form type-C precipitates [21] and the Q phase [22]. As a result, in this study, the T6 sample mainly contained the *Q* phase and Mg_2Si precipitates after aging. The present XRD spectra did not detect the *S*-Al_2CuMg phase from the T6 sample. Neither could Vieira find the *S*-Al_2CuMg phase in his study, in which two age-hardening heat treatments of Al–10Si–4.5Cu–2Mg were completed [23].

The CFT6a sample increased the HAGBs in its matrix to provide more potent sites for accommodating Cu atoms after the solution treatment. During quench or aging in room temperature, Cu diffused to tie up with Al forming Cu_9Al_4 phase. During artificial aging, a Mg–Si cluster formed *in situ*; this either led to the formation of CuMgSi precipitates through movement of the Cu atoms, or the Cu–Mg clusters formed first and subsequently incubated the CuMgSi precipitate in the matrix. Figure 5a,b shows the transmission electron microscopy photos of the T6 and CFT6a samples, respectively. For a given solution and aging conditions, the CFT6a sample achieved finer precipitates (less than 100 nm) than the T6 sample did. The main constituted phases likely included Al(Mn,Fe)Si in block and plate shapes, and some fine particles (less than 100 nm) likely being Cu_9Al_4 and CuMgSi precipitates or Cu atoms surrounding fine Mg_2Si phases, as shown in Figure 5b. The α-Al(Mn,Fe)Si dispersoids could have a block-shaped or plate-shaped morphology in the size of 50–200 nm, as reported by Li *et al.* [24].

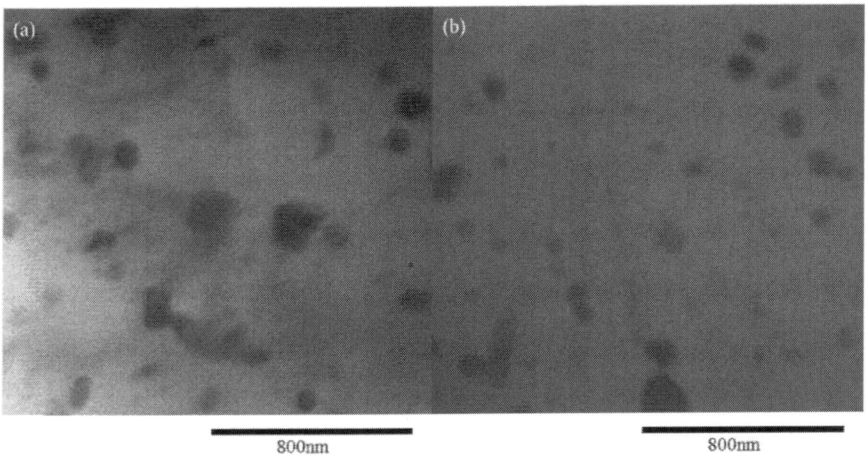

Figure 5. TEM photo show the plate-shape and lump-shape precipitate in the matrix of (**a**) T6 sample, and (**b**) CFT6a sample.

An electron probe X-ray microanalyzer (JXA-8200, JEOL USA, Inc., Peabody, MA, USA) was used to obtain images of Mg, Si, Mn, and Cu mappings from the matrix of the CFT6a sample. The white aggregates in Figure 6a are the Al(Mn,Fe)Si and*Q* phase, and the black lumps are Mg_2Si particles. The arrows indicate the locations of *Q*-phase precipitates. The CuMgSi and Cu_9Al_4 precipitates are nanoscale and were difficult to detect by mapping. In Figure 6b, the white particles shown in the matrix of the T6 samples are Al(Mn,Fe)Si and *Q*-AlCuMgSi phases. An Al(MnFe)Si particle attached with Cu atoms could be distinguished and are indicated by an arrow. The black spots are the Mg_2Si phase.

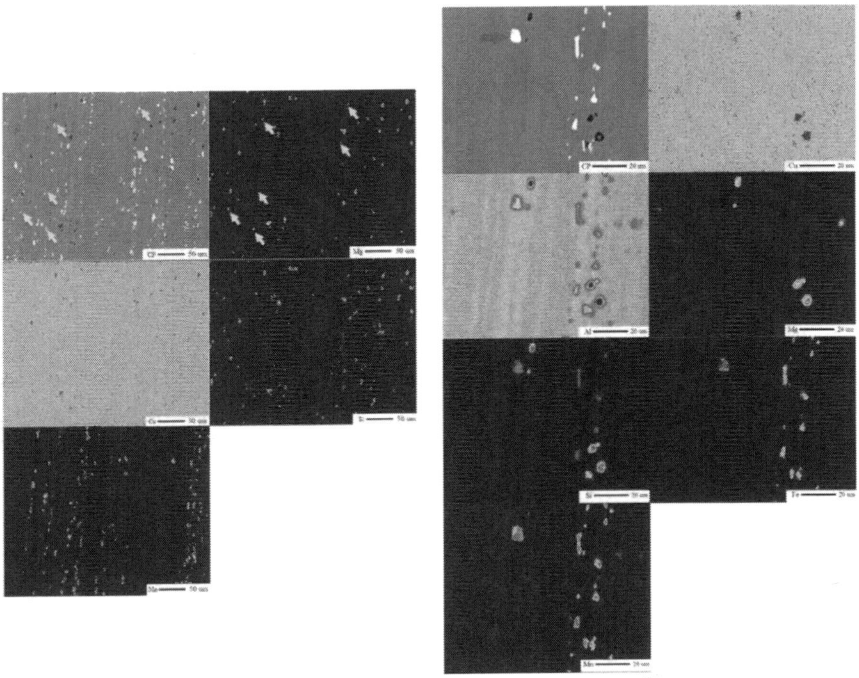

(a) (b)

Figure 6. SEM photo and alloying elements mappings obtained from (**a**) T6 sample and (**b**) CFT6a sample.

The *Q* phase, Al(Mn,Fe)Si, and Mg_2Si (larger than 80 nm) contained in the aluminum alloy sample are non-shearable particles in the α-Al matrix [25]. Therefore, the T6 samples gained high strength. By contrast, the CFT6a samples contained Al(Mn,Fe)Si, Mg_2Si, fine CuMgSi, and Cu_9Al_4 precipitates. The fine Mg_2Si or CuMgSi and Cu_9Al_4 precipitates are shearable.Figure 7 illustrates the dislocations (marked as 1-1 and 2-2) intersected with two fine precipitates. In addition, the SP/IMC particle counts (Figure 2) were lower in the matrix of the CFT6a samples. As a result, the CFT6a sample had decreased numbers of barrier sites for tangling dislocations to reduce strength but enhance elongation.

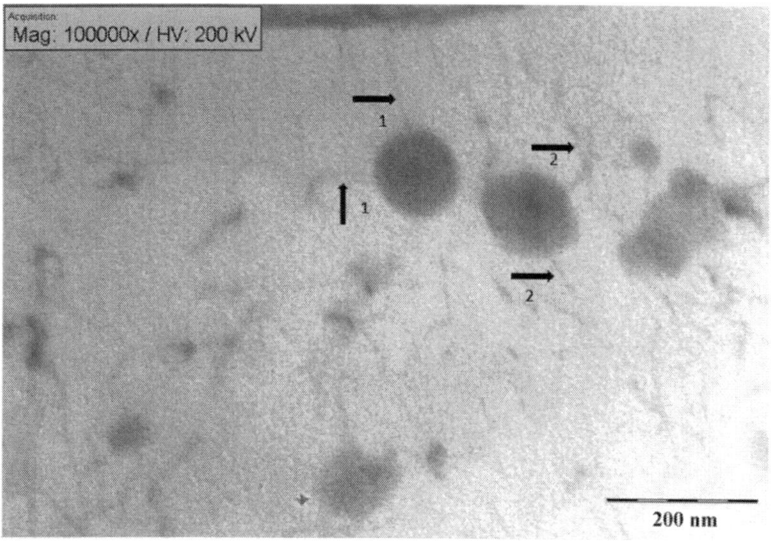

Figure 7. TEM photo shows the intersection of dislocation 1-1 and 2-2 with two fine precipitates in the matrix of CFT6a sample, respectively.

Table 3 compares the surface roughness of T6 and CFT6a samples before and after anodization/sealing. Compared with the samples without anodization, these anodized and sealed samples showed an approximately 8%–10% reduction in elongation. Such a decrease is likely caused by the dissolution of SP/IMC at the film/matrix interface leading to degrade surface roughness and elongation. Before anodization, the surface roughness of the T6 and CFT6a sample was approximately 0.07 (0.04) μm; after anodization and sealing, the surface roughness became 0.16 (0.03) and 0.13 (0.06) μm.

Table 3. Measured surface roughness of different samples before and after anodization.

Sample	Surface Roughness (μm)
T6; CFT6a	0.07 (0.04)
T6-A	0.16 (0.03)
CFT6a-A	0.13 (0.06)

Note: the deviations are listed in parentheses.

3.2. Fatigue and Corrosion Tests

Applying the cryoforge before the solution treatment reduced the SP/IMC particle counts and transformed the *Q* phase into nanoscale CuMgSi and Cu_9Al_4 precipitates, which reduced UTS and YS but increased elongation. Figure 8 shows that both the T6 and CFT6a samples obtained fatigue strength of 180 MPa at 1×10^7 life cycles. As shown in Figure 3a,c, we found that cyclic loading functioned to shift the peak of relative frequency from the misorientation angle of 40°–45° to 45°–50°, and increase some grain boundaries at angles of 10°–20°. During cyclic loading, shear stress drove part of the dislocations to conduct polygonization and/or reorganization.

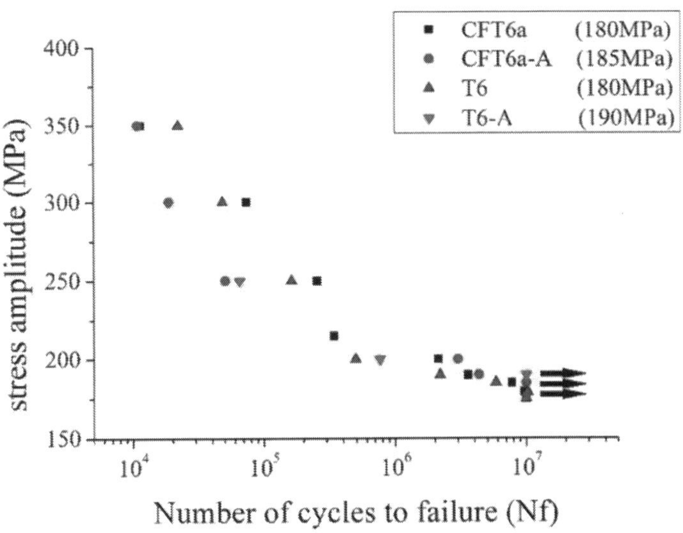

Figure 8. S–N curves for different samples; including T6 and CFT6a with/without anodization/sealed treatment; "A" represented anodization/ sealed sample.

Figure 9a,b show the fractured surface of the T6 and CFT6a samples, both of which were subjected to 185 MPa and which performed 5.8×10^6 and 7.7×10^6 life cycles, respectively. The CFT6a sample exhibited narrower striation spacing than did the T6 samples and achieved a longer fatigue life. The T6 sample obtained fracture steps, as revealed in Figure 9a, which can be attributed to the *Q* phase located at or near grain boundaries that served to strengthen the grain boundaries [22,26]. Increasing fine precipitates in the matrix of the CFT6a sample likely more effectively to consume the crack propagation energy and thus slightly enhanced the life cycles.

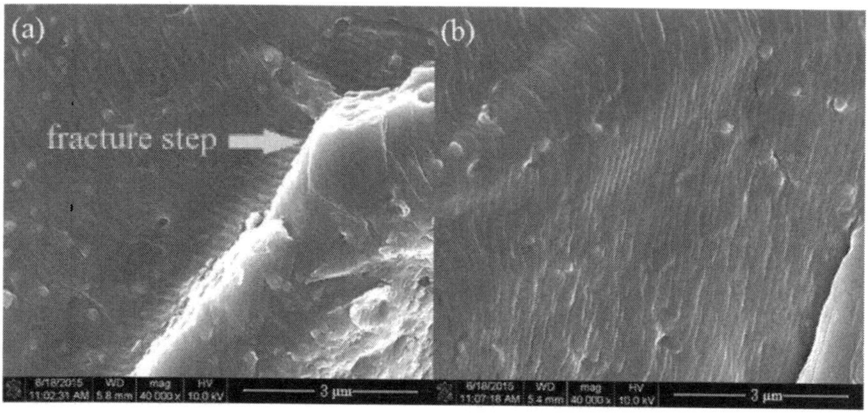

Figure 9. Fracture surface prepared from (**a**) T6; (**b**) CFT6a samples; both subjected to 185 MPa and fractured at 5.8×10^6 and 7.7×10^6 life cycles, respectively.

The SP/IMC particles were small to be less than 17 μm. Anodization did not yield the deteriorated effect of reducing fatigue strength. The two anodized and sealed samples (T6-A and CFT6a-A) showed similar fatigue strength (185–190 MPa) at 1×10^7 life cycles as did those without anodization and sealing (180 MPa).

The polarization curves for all the samples with and without sealed anodic aluminum oxide (AAO) films are shown inFigure 10. The corrosion behavior of the samples was affected by the constituent phases of each. The T6 samples contained AlCuMgSi, Al(FeMn)Si, and Mg_2Si phases, whereas the CFT6a samples acquired mainly Al(FeMn)Si and Mg_2Si phases as well as some CuMgSi and Cu_9Al_4 precipitates. Mg_2Si is anodic relative to the aluminum matrix, but Al(FeMn)Si and AlCuMgSi are cathodic. The Cu–Al precipitates are more vulnerable to attack during the immersion test, compared with the$Q$ phase [23]. The CFT6a samples exhibited slightly inferior E_{corr} (−1.0 V) and I_{corr} (4.7×10^{-6} A/cm^2) than the T6 sample did (−0.93 V and 2.6×10^{-6} A/cm^2, respectively).

After anodization and sealing, the anodized films on the aluminum alloy samples contained mainly amorphous alumina and a few hydrated alumina [27]. These AAO films could significantly enhance the corrosion resistance of the aluminum alloys [28]. Therefore, the anodized films deposited on the T6 and CFT6 samples improved E_{corr} from (−0.9 to −1.0 V) to (−0.64 to −0.76 V).

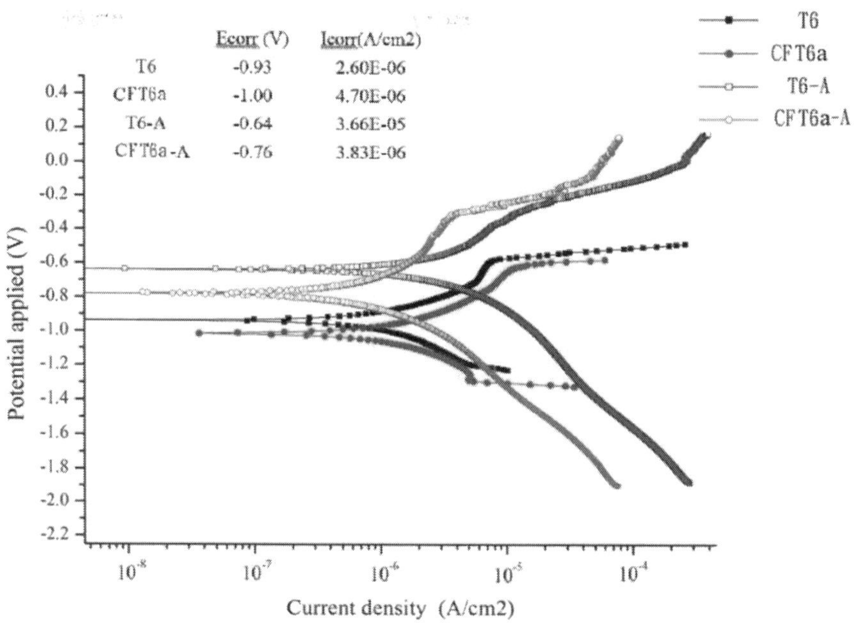

Figure 10. Polarization curves obtained from immersion tests of T6 and CFT6a samples with/without anodization/sealing treatment.

Affected by a greater number of coarse SP/IMC (Table 2) and/or a higher particle population, the anodized and sealed film on the T6 sample was entrapped with particles along with air pockets, as shown in Figure 11a. By contrast, the film on the CFT6a sample is relatively sound with few trapped particles; see Figure 11b. The trapped particles could function as corrosion channels to accelerate chloride attacks. Therefore, the anodized and sealed T6 sample obtained a higher current density (I_{corr} of 3.6 × 10^{-5} A/cm^2) than did the anodized and sealed CFT6a sample (I_{corr} of 3.8 × 10^{-6} A/cm^2).

The dissolution of SP/IMC at the film–matrix interface drove aluminum and magnesium ions to move toward electrolytes leaving the Si particles remaining in the film, as shown in Figure 11c [27]. Consequently, the anodized and sealed T6 sample obtained inferior corrosion current density to the bare T6 sample (3.6 × 10^{-5} A/cm^2 *vs.* 2.6 × 10^{-6} A/cm^2). Decreasing SP/IMC particle size and counts also reduced the size and counts of the Si particles that remained in the anodized film to undergo a minor change in corrosion current density (4.7 × 10^{-6} *vs.* 3.86 × 10^{-6} A/cm^2) in the bare CFT6a and anodized/sealed CFT6a samples.

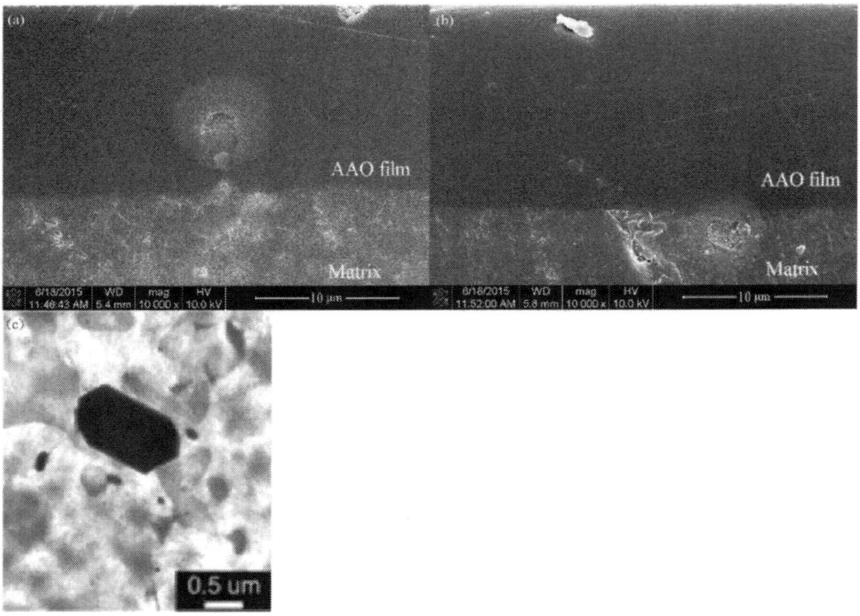

Figure 11. SEM photos show the cross-section of deposited anodized/sealed films on (**a**) T6 sample; (**b**) CFT6a sample; and (**c**) anodized/sealed 6063-T6 samples [27].

4. CONCLUSIONS

Adding cryoforging to the process changed the microstructure and mechanical properties of the AA6066–T6 alloy samples; specifically, it decreased SP/IMC particle counts and reduced the *Q* phase but increased fine CuMgSi and Cu_9Al_4precipitates in the matrix of the CFT6a sample. As a result, the tensile property of elongation was increased by approximately 34% comparing with T6 sample. The anodized and sealed AAO film on the CFT6a sample improved its corrosion resistance but decreased its elongation by approximately 10% (from 14.2% to 12.7%).

ACKNOWLEDGMENTS

We gratefully acknowledge the financial support from the Ministry of Science and Technology of the Republic of China (MOST 103-2221-E-008-026-MY2). Many thanks also to National Central University for providing the SEM and TEM tests and to National Sun-Yat Sen University for the EBSD analysis.

AUTHOR CONTRIBUTIONS

Hwa-Sheng Yong and Wen-Nong Hsu ran experiments of this study and did OM, SEM, TEM observation and EBSD analyses. Main contribution also included experimental data collection. All authors provided equal contribution.

CONFLICTS OF INTEREST

The authors declare no conflict of interest.

REFERENCES

1. Wang, Y.M.; Ma, E.; Chen, M.W. Enhanced tensile ductility and toughness in nanostructured Cu. *Appl. Phys. Lett.* **2002**, *80*, 2395–2397.

2. Rangaraju, N.; Raghuram, T.; Krishna, B.V.; Rao, K.P.; Venugopal, P. Effect of cryo-rolling and annealing on microstructure and properties of commercially pure aluminium. *Mater. Sci. Eng. A Struct.* **2005**, *398*, 246–251.

3. Chen, Y.J.; Roven, H.J.; Gireesh, S.S.; Skaret, P.C.; Hjelen, J. Quantitative study of grain refinement in Al–Mg alloy processed by equal channel angular pressing at cryogenic temperature. *Mater. Lett.* **2011**, *65*, 3472–3475.

4. Lee, Y.B.; Shin, D.H.; Park, K.T.; Nam, W.J. Effect of annealing temperature on microstructures and mechanical properties of a 5083 Al alloy deformed at cryogenic temperature. *Scr. Mater.* **2004**, *51*, 355–359.

5. Panigrahi, S.K.; Jayagathan, R. A study on the mechanical properties of cryorolled Al–Mg–Si alloy. *Mater. Sci. Eng. A Struct.* **2008**, *480*, 299–305.

6. Panigrahi, S.K.; Jayagathan, R.; Chawla, V. Effect of cryorolling on microstructure of Al–Mg–Si alloy. *Mater. Lett.* **2008**,*62*, 2626–2639.

7. Panigrahi, S.K.; Jayagathan, R. Development of ultrafine-grained Al 6063 alloy by cryorolling with the optimized initial heat treatment conditions. *Mater. Des.* **2011**, *32*, 2172–2180.

8. Krishna, K.G.; Sivaprasad, K.; Venkateswarlu, K.; Kumar, K.C.H. Microstructural evolution and aging behavior of cryorolled Al–4Zn–2Mg alloy. *Mater. Sci. Eng. A Struct.* **2012**, *535*, 129–135.

9. Das, P.; Jayaganthan, R.; Singh, I.V. Tensile and impact-toughness behavior of cryorolled Al 7075 alloy. *Mater. Des.***2011**, *32*, 1298–1305.

10. Panigrahi, S.K.; Jayaganthan, R. Effect of ageing on microstructure and mechanical properties of bulk, cryorolled, and room temperature rolled Al 7075 alloy. *J. Alloy. Compd.* **2011**, *509*, 9609–9016.

11. Yin, J.; Lu, J.; Ma, H.; Zhang, P. Nanostructural formation of fine grained aluminum alloy by severe plastic deformation at cryogenic temperature. *J. Mater. Sci.* **2004**, *39*, 2851–2854.

12. Krishna, K.G.; Sivaprasad, K.; Narayanana, T.S.N.S.; Kumar, K.C. Localized corrosion of an ultrafine grained Al–4Zn–2Mg alloy produced by cryorolling. *Corros. Sci.* **2012**, *60*, 82–89.

13. Shanmugasundaram, T.; Murty, B.S.; Sarma, V.S. Development of ultrafine grained high strength Al–Cu alloy by cryorolling. *Scr. Mater.* **2006**, *54*, 2013–2017.

14. Rao, P.N.; Jayaganthan, R. Effects of warm rolling and ageing after cryogenic rolling on mechanical properties and microstructure of Al 6061 alloy. *Mater. Des.* **2012**, *39*, 226–233.

15. Huang, Y.S.; Shih, T.S.; Chou, J.H. Electrochemical behavior of anodized AA7075-T73 alloys affected by matrix structures. *Appl. Surf. Sci.* **2013**, *283*, 249–257.

16. *ASTM B557-15, Standard Test Methods for Tension Testing Wrought and Cast Aluminum- and Magnesium-Alloy Products*; ASTM International: West Conshohocken, PA, USA, 2015.

17. *JIS Z2274, Method of Rotating Bending Fatigue Testing of Metals*; Japanese Standards Association: Tokyo, Japan, 1978.

18. Shih, T.S.; Lee, T.H.; Jhou, Y.J. The effects of anodization treatment on the microstructure and fatigue behavior of 7075-T73 aluminum alloy. *Mater. Trans.* **2014**, *55*, 1280–1285.

19. Kim, H.J.; Lee, J.Y.; Paik, K.W.; Koh, K.W.; Won, J.; Choe, S.; Lee, J.; Moon, J.T.; Park, Y.J. Effects of Cu/Al intermetallic compound (IMC) on copper wire and aluminum pad bondability. *IEEE Trans. Pack. Technol.* **2003**, *26*, 367–374.

20. Du, Y.; Chang, Y.A.; Huang, B.; Gong, W.; Jin, Z.; Xu, H.; Yuan, Z.; Liu, Y.; He, Y.; Xie, F.Y. Diffusion coefficients of some solutes in

fcc and liquid Al: Critical evaluation and correlation. *Mater. Sci. Eng. A Struct.* **2003**, *363*, 140–151.

21. Matsuda, K.; Teguri, D.; Sato, T.; Uetani, Y.; Ikeno, S. Cu Segregation around Metastable Phase in Al–Mg–Si Alloy with Cu. *Mater. Trans.* **2007**, *48*, 967–974.

22. Svenningsen, G.; Larsen, M.H.; Walmsley, J.C.; Nordlien, J.H.; Nisancioglu, K. Effect of artificial aging on intergranular corrosion of extruded AlMgSi alloy with small Cu content. *Corros. Sci.* **2006**, *48*, 1528–1543.

23. Vieira, A.C.; Pinto, A.M.; Rocha, L.A.; Mischler, S. Effect of Al_2Cu precipitates size and mass transport on the polarisation behaviour of age-hardened Al–Si–Cu–Mg alloys in 0.05 M NaCl. *Electrochim. Acta* **2011**, *56*, 3821–3828.

24. Li, Y.J.; Muggerud, A.M.F.; Olsen, A.; Furu, T. Precipitation of partially coherent α-Al(Mn,Fe)Si dispersoids and their strengthening effect in AA 3003 alloy. *Acta Mater.* **2012**, *60*, 1004–1014.

25. Cabibbo, M. Microstructure strengthening mechanisms in different equal channel angular pressed aluminum alloys.*Mater. Sci. Eng. A Struct.* **2013**, *560*, 413–432.

26. Zhaia, T.; Jianga, X.P.; Lia, J.X.; Garratt, M.; Bray, G.H. The grain boundary geometry for optimum resistance to growth of short fatigue cracks in high strength Al-alloys. *Int. J. Fatigue* **2005**, *27*, 1202–1209.

27. Huang, Y.S.; Shih, T.S.; Wu, C.E. Electrochemical behavior of anodized AA6063-T6 alloys affected by matrix structures. *Appl. Surf. Sci.* **2013**, *264*, 410–418.

28. Shih, T.S.; Chiu, Y.W. Corrosion resistance and high-cycle fatigue strength of anodized/sealed AA7050 and AA7075 alloys. *Appl. Surf. Sci.* **2015**, *351*, 997–1003.

CHAPTER 6

Effects of Deep Rolling and Its Modification on Fatigue Performance of Aluminium Alloy AA6110

Patiphan Juijerm[1] and Igor Altenberger[2]

[1] Materials Innovation Center, Department of Materials Engineering, Kasetsart University,, Thailand

[2] WIELAND-WERKE AG, Central Laboratory, Research & Development,, Germany

1. INTRODUCTION

Recently, low-weight components are particularly required for environmental, ecological and economical aspects. Therefore, light-weight metals are frequently mentioned and selected for many applications where low density and high strength to weight ratios are an important consideration. Consequently, development and improvement in the field of light-weight metals can be seen continuously for advanced applications in automotive as well as aerospace industries. One of the most important light-weight metals is aluminium and its alloys which possess many attractive characteristics including excellent corrosion resistance in most environments, reflectivity, high strength and stiffness to weight ratio, good formability, weldability and recycling potential [1,2]. Certainly, these advantageous properties make them ideal candidates to replace heavier materials (steel or copper) for several industries. Therefore, mechanical behavior of aluminium alloys becomes more and more important, especially under cyclic loading due to failures occurring in machinery components are almost entirely fatigue failures. Accordingly, fatigue performance of aluminium alloys was investigated and also improved by mechanical surface treatments, e.g. shot peening, deep rolling and laser shock peening. Deep rolling is one of the most well-known mechanical surface treatment methods and exhibits a great depth of near-surface work hardening state and compressive residual stresses serving to inhibit or retard fatigue crack initiation as well as crack growth [3-5].

Ageing treatments combined with e.g. shot peening or deep rolling are methods to enhance the effects of conventional mechanical surface treatments for steels or materials having interstitial solute atoms. The fatigue performance of deep rolled steels SAE 1045, AISI 4140 and stainless steel AISI 304 can be considerably improved after annealing due to static/dynamic strain ageing. The formation of carbon atom clouds and very small carbide precipitates are the significant phenomena to impede dislocation movement and thus determine the mechanical properties. Normally, for precipitation-hardenable materials such as aluminium- and titanium alloys, mechanical surface treatment is performed after ageing treatments for fatigue lifetime enhancement. However, in some cases, mechanical surface treatment is performed on solution-heat-treated conditions with subsequent ageing treatments to produce increased hardness by precipitates especially in near-surface regions. The fine precipitates can be reasonably produced by prior plastic deformation due to preferential heterogeneous nucleation in the vicinity of dislocations in near-surface regions thus improving the cyclic deformation behavior [6,7]. On the other hand, the near-surface compressive residual stresses and work hardening induced by mechanical surface treatments decrease during annealing as well as ageing treatments. By the reduction of compressive residual stresses and near-surface work hardening, a detrimental effect for the fatigue lifetime is expected. The effects of decreased residual stress as well as work hardening together with the increased hardness by the ageing treatment on the fatigue performance are of particular interest, especially in smooth, soft and mechanically surface treated aluminium alloys because their fatigue lifetime depends significantly on the stability of near-surface work hardening as well as compressive residual stresses which can inhibit or retard surface fatigue crack initiation as well as fatigue crack growth. The purpose of this work is therefore to investigate systematically how optimized fatigue lifetimes by ageing treatments of deep rolled as-quenched AA6110 can be obtained. It should be noted that the optimize ageing treatment for the solution-heat-treated aluminium alloy AA6110 producing a maximum hardness value and also maximum yield stress is at a temperature of 160 °C for about 12 hr. Moreover, modern mechanical surface treatments have been developed from shot peening and deep rolling to give an optimized surface condition. At elevated temperatures, static/dynamic strain ageing is applied to shot peening and deep rolling. Consequently, warm shot peening and high-temperature deep rolling have become established techniques, particularly for materials that have interstitial solute atoms, e.g. steels containing carbon or nitrogen atoms [7]. Superior stability of macroscopic compressive residual stresses as well as work-hardening states during cyclic loading are observed due to dislocation pinning by interstitial solute atoms (so-called Cottrell clouds) or very fine carbides. Accordingly, superior fatigue performance was observed as compared to conventional mechanical surface treatments. It is very challenging when a modern mechanical surface treatment such as high temperature deep rolling is performed on precipitation-hardenable materials such as aluminium alloys that have mainly substitutional solute atoms where the full beneficial effects of static/dynamic strain ageing cannot be expected. Nevertheless, static/dynamic precipitation occurring during mechanical surface treatment at elevated temperatures may contribute to mechanical properties of the surface as well as the bulk, particularly for the solution-heat-treated condition. Thus, a newly developed mechanical surface treatment, high-temperature deep rolling, was performed on the solution-heat-treated aluminium alloy AA6110 for different deep-rolling temperatures up to 250 °C.

2. METHODOLOGY

The aluminium wrought alloy AA6110 was delivered from Alcoa Extrusions, Hannover, Germany as extruded bars with a diameter of 34 mm. The chemical composition of this alloy is 0.86 Si, 0.19 Fe, 0.45 Cu, 0.46 Mn, 0.78 Mg, 0.17 Cr, 0.02 Zn, 0.01 Ti and Al balance (all values in wt%). The alloy is a precipitation-hardenable Al–Mg–Si–Cu alloy, forming ´´ as well as Q´ precipitates. Cylindrical specimens with a diameter of 7 mm and a gauge length of 15 mm were prepared. The loading direction during fatigue investigations corresponds to the extrusion direction of the bar. The specimens were solution heat treated in a furnace with argon atmosphere at a temperature of 525 °C for 30 min followed by water quenching to room temperature. The specimens were deep rolled at room temperature immediately after quenching. A hydraulic rolling device with a 6.6 mm spherical rolling element and a rolling pressure of 80 bar was used. The deep rolled as-quenched specimens were aged in the temperature range 50–300 °C with different ageing times up to approximately 28 and 148 h. For a high-temperature deep rolling, a pneumatic rolling device with a 40 mm diameter roller and a rolling force of 0.27 kN was applied at room temperature and at elevated temperatures up to 250 °C using inductive heating. On all specimens fatigue tests as well as thermal relaxation investigations were carried out. The non-mechanically surface treated specimens were electrolytically polished in the gauge length leading to a material removal of 100 μm before testing to avoid any influence of machining. Tension-compression fatigue tests were conducted with a servohydraulical testing device at stress control without mean stress (R = -1) and with a test frequency of 5 Hz. Strain was measured using capacitative extensometers. Residual stress- and FWHM-value-depth-profiles were determined by successive electrolytical material removal using the classical $\sin^2\Psi$-method with CuKα-radiation and the {333}-planes and 1/2 s_2 = 19.77 x 10^5 mm^2/N as elastic constant. All residual stresses and FWHM-values were measured in longitudinal direction of the specimens. No stress correction was carried out after electrolytical material removal of surface layers.

3. RESULTS AND DISCUSSION

3.1. Deep Rolling Followed By Ageing Treatment

Normally, for precipitation-hardenable materials, such as aluminium- and titanium alloys, mechanical surface treatment is performed after ageing treatments for fatigue lifetime enhancement as reported in [5,8-11]. However, in some cases, mechanical surface treatment is performed on solution heat treated (as-quenched) conditions with subsequent ageing treatments to produce increased hardness by precipitates especially in near-surface regions [12-15]. However, on the other hand, at the same time, near-surface macroscopic compressive residual stresses and work hardening states induced by mechanical surface treatments decrease during ageing treatments due to thermal residual stress relaxation. By the reduction of near-surface macroscopic compressive residual stresses and work hardening states, a detrimental effect for the fatigue lifetime can be expected. The effects of decreased residual stresses as well as work hardening states together with increased hardness values by the ageing treatment on the fatigue behavior of aluminium alloy AA6110 will therefore be thoroughly investigated and closely monitored in this section. Fig. 1 shows the hardness values at the surface of the deep rolled as-quenched condition after ageing treatments as a function of ageing times and temperatures. After ageing treatments in the temperature range 160-250 °C, the hardness values at the surface increased continuously with increasing ageing time until reaching a maximum value. Maximum hardness values of approximately 140, 133 and 123 HV were measured after ageing at temperatures of 160 200 and 250 °C for ageing times of approximately 12 hr, 1,000 and 100 seconds, respectively. For an ageing temperature of 300 °C, increased hardness values at the surface were not observed. The precipitated phases, " as well as Q' lead to the increased hardness of copper-containing Al-Mg-Si aluminium alloys [1,2]. The maximum hardness of the deep rolled as-quenched AA6110 can be found after an ageing treatment at a temperature of 160 °C and an ageing time of 12 hr. For prolonged ageing treatments in the temperature range of 160-250 °C, the hardness values at the surface of deep rolled as-quenched specimens declined after having reached the peak hardness. The formation of coarse, semi-coherent, ' and/or Q' as well as incoherent precipitates, and/or Q at the surface as well as in near-surface regions is the reason for this observation. Conversely, for an ageing treatment at a temperature of 300 °C with a short ageing time, an increase of hardness was not observed. It might be due to the dominant recrystallization process taking place before occurring precipitation process [16].

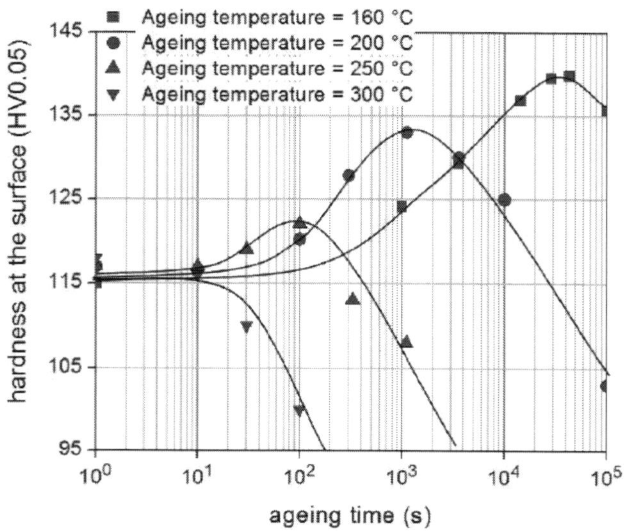

Figure 1. The hardness values at the surface of deep rolled as-quenched AA6110 during different ageing treatments.

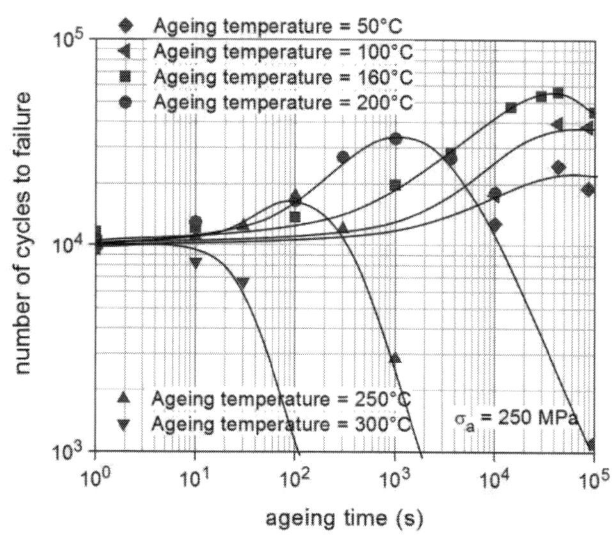

Figure 2. Fatigue lifetime as a function of ageing time and –temperature for an applied stress amplitude of 250 MPa

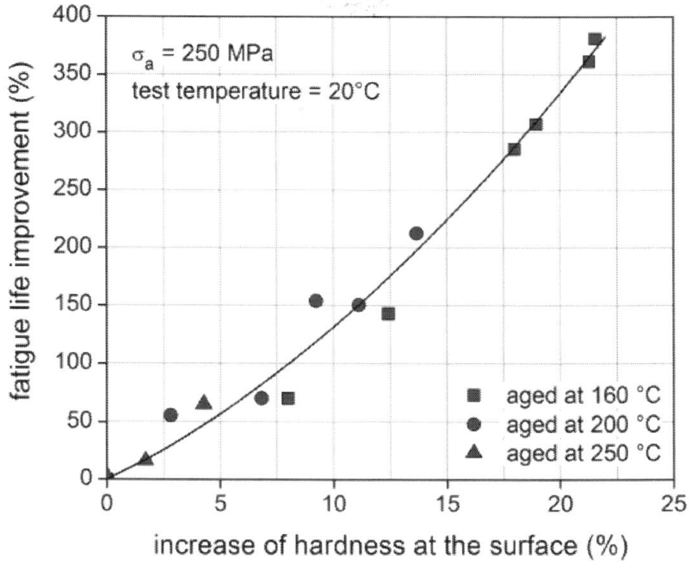

Figure 3. Fatigue life improvement as a function of hardness increase at the surface of deep rolled as-quenched AA6110 after ageing treatments at 160-250 °C.

Fatigue tests were also performed to evaluate the optimized fatigue lifetime. For this purpose, the relation between ageing time, temperature and fatigue lifetime was established at an applied stress amplitude of 250 MPa as shown in Fig. 2. After ageing treatments in the temperature range 160-250 °C, the fatigue lifetimes increased continuously with increasing ageing time until reaching a maximum lifetime. The optimized fatigue lifetime of the deep rolled as-quenched AA6110 was found after an ageing treatment at a temperature of 160 °C for about 12 hr. As expected, the increase of near-surface hardness after the ageing treatments resulted in an enhancement of the fatigue lifetime. Fig. 1 and Fig. 2 show that there is a clear correlation between hardness values and fatigue lifetimes of the deep rolled as-quenched AA6110 after ageing treatments. To clarify this correlation, a diagram of fatigue life improvement versus increase of hardness at the surface was constructed as shown in Fig. 3 using experimental data from Fig. 1 and Fig. 2. The residual stresses and FWHM-values decreased due to the relaxation as well as recovery processes. Residual stress and FWHM-values at the surface were reduced from -265 to -100 MPa and from 2.3° to 2.0°, respectively. Because residual stress relaxation immediately at the surface is stronger than in subsurface layers, after ageing, a subsurface compressive residual stress maximum is formed as shown in Fig. 4. Non-statistically evaluated s/n-curves of the deep rolled as-quenched state after the optimized ageing treatment are shown as compared to the deep rolled as-quenched conditions in Fig. 5. At room temperature, the ageing treatment after mechanical surface treatment can enhance the fatigue lifetime especially in low cycle fatigue regime. However, in the high cycle fatigue regime, the difference in fatigue lifetimes was almost negligible. For elevated test temperature, the optimized-aged deep rolled condition exhibits also greater fatigue lifetimes as compared to the deep rolled as-quenched condition. Fatigue lifetimes of the deep rolled as-quenched condition were improved especially in the low cycle fatigue regime after the optimized ageing treatment (see Fig. 5) due to increased hardness values at the surface and in near-surface regions as well as in the bulk. However, the induced macroscopic compressive residual stresses as well as work hardening states seem to be essential in the high cycle fatigue regime, where relatively low stress amplitudes were applied and mechanical relaxation was not significant during cyclic loading at room temperature.

3.2. Comparison Of Conventional And Modified Deep Rolling

The most important and interesting issue of the modified mechanical surface treatment (deep rolling followed by optimized ageing treatment) is the comparison with the conventional mechanical surface treatment (optimized ageing followed by deep rolling). First of all, important information of surface properties of the optimized-aged deep rolled as-quenched and deep rolled optimized condition is shown in Fig. 6 illustrating residual stress- and FWHM-depth-profiles of deep rolled as-quenched AA6110 after the optimized ageing treatment as compared to deep rolled optimized-aged AA6110. Hardness values of 137 and 161 HV are measured at the surface of deep rolled as-quenched AA6110 after the optimized ageing treatment and deep rolled optimized-aged AA6110, respectively. Noticeably, all important properties which are beneficial effects for fatigue lifetime enhancement, such as hardness, macroscopic compressive residual stress as well as the work hardening state of the optimized-aged deep rolled as-quenched condition are significantly less than of the deep rolled optimized/peak-aged condition. The greater near-surface macroscopic compressive residual stresses, FWHM-values as well as hardness values of the deep rolled optimized condition indicate that the deep rolling after an optimized ageing treatment results in an excellent combination of work- and precipitation hardening and thus excellent fatigue lifetime could be expected for the deep rolled optimized/peak-aged AA6110. To confirm this assumption, s/n-curves of the optimized-aged deep rolled as-quenched condition were plotted and compared to the deep rolled optimized/peak-aged condition in one diagram in Fig. 7. Obviously, fatigue lifetimes as well as -strength of the deep rolled optimized/peak-aged condition are superior to the optimized-aged deep rolled as-quenched condition. It can be concluded that both hardening effects, work- and precipitation hardening are required to yield the best fatigue lifetimes of AA6110. As known, the deep rolling treatment serves principally to induce near-surface work hardening and macroscopic compressive residual stress. For the ageing treatment after deep rolling, unfortunately, the work hardening and macroscopic compressive residual stresses were partially annealed out rapidly during the ageing treatment due to the relaxation process (mainly self diffusion). Therefore it can reasonably be assumed that fatigue lifetimes of the optimized-aged deep rolled as-quenched AA6110 were governed by the precipitation hardening and residually effective work hardening as well as compressive residual stresses at the surface and in near-surface regions. Therefore, the fatigue lifetime of the deep rolled as-quenched AA6110 followed by the optimized ageing treatment is better than of the deep rolled as-quenched AA6110 in the low cycle fatigue regime (seeFig. 5). However, the deep rolling after a suitable/optimized ageing treatment can completely combine the work- and precipitation hardening in near-surface regions of AA6110 into an optimized microstructure and thus result in the best surface properties and fatigue lifetime of the investigated AA6110 (see Fig. 7).

3.3. Deep Rolling At Elevated Temperature

Thermomechanical surface treatment, the high-temperature deep rolling, has been successfully investigated for SAE 1045 as well as AISI 304 [20-24]. The fatigue behavior of these steels can be enhanced considerably due to static/dynamic strain ageing and together with very fine carbides at the surface and in near-surface regions [20-24]. Nevertheless, for aluminium alloys, it is still doubtful whether thermomechanical surface treatments can enhance the fatigue behavior more significantly than conventional mechanical surface treatments because aluminium alloys have mainly substitutional solute atoms. Thus, the fully beneficial effects of static/dynamic strain ageing can not be expected. However, static/dynamic precipitation during mechanical surface treatment at elevated temperature may contribute to mechanical properties of the surface as well as the bulk particularly for the as-quenched condition. Therefore, high-temperature deep rolling on the as-quenched aluminium alloy AA6110 was investigated. As-quenched specimens were deep rolled at different elevated temperatures of 160, 200 and 250 °C. Afterwards, near-surface properties and fatigue behavior were investigated and presented in this section. After deep rolling at elevated temperatures, near-surface residual stress, work hardening state and hardness-depth-profiles were measured as compared to the room-temperature deep rolled state as shown in Fig. 8 and Fig. 9. Obviously, macroscopic compressive residual stresses tend to decrease with increasing deep rolling temperature. Maximum macroscopic compressive residual stresses of -181, -152 and -59 MPa were measured at a depth of 20 μm after deep rolling at temperatures of 160, 200 and 250 °C, respectively. In contrast, after deep rolling at room temperature, a maximum macroscopic compressive residual stress value of -286 MPa was measured directly at the surface (see Fig. 6). After deep rolling at a temperature of 160 °C an approximately FWHM-value of 2.3° was measured which was to identical to the one observed after room-temperature deep rolling. However, FWHM-values tend to decrease at high temperature with increasing deep rolling temperature. The FWHM-values about 2.1 and 1.6° were detected after deep rolling at temperatures of 200 and 250 °C, respectively. In addition, the case depth of work hardening after deep rolling at elevated temperatures seem to be greater than after room-temperature deep rolling. Near-surface hardness values, however, increased with increasing deep rolling temperature up to 200 °C as compared to deep rolling at room temperature. Hardness values in a depth of 25 μm of about 125 and 134.5 HV were measured after deep rolling at temperatures of 160 and 200 °C, respectively, whereas after deep rolling at room temperature, the hardness in a depth of 25 μm was approximately 113 HV. Conversely, after deep rolling at a temperature of 250 °C, a hardness in a depth of 25 μm of only about 104 HV was observed. As expected, due to occurring static/dynamic precipitation during deep rolling at elevated temperatures, near-surface hardness values after high-temperature deep rolling (160-200 °C) increased as compared to the room-temperature deep rolled as-quenched AA6110 (see Fig. 9). On the other hand, lower macroscopic compressive residual stresses and work hardening states were measured as compared to the room-temperature deep rolled as-quenched condition (see Fig. 8)because static/dynamic recovery processes, which bring about relaxation phenomena, took place during deep rolling at elevated temperatures. The deep rolling treatment at a temperature of 250 °C produced detrimental effects on the near-surface properties, i.e. near-surface macroscopic compressive residual stresses, work hardening states and hardness values are considerably lower than of the room-temperature deep rolled as-quenched condition. That might be due to the fact that this temperature too high for the aluminium alloy AA6110 and leads to serve over-ageing effects and a high-rate static/dynamic recovery for this situation.

Non-statistically evaluated s/n-curves of the differently high-temperature deep rolled as-quenched condition at room temperature are presented as compared to the room-temperature deep rolled as-quenched condition in Fig. 10. The difference fatigue lifetimes for deep rolling treatments at temperatures between room temperature and 200 °C were insignificant. In the low cycle fatigue regime, fatigue lifetimes of as-quenched specimens deep rolled at a temperature of 200 °C seem to be slightly better than of the room-temperature deep rolled as-quenched condition, however in the high cycle fatigue regime, a contrary behavior was seen. The improvement of fatigue lifetimes at room temperature of the high-temperature deep rolled as-quenched AA6110 is not obvious as compared to the room-temperature deep rolled as-quenched AA6110. However, increased near-surface hardness values after deep rolling at a temperature of 200 °C slightly enhance fatigue lifetimes at room temperature in the low cycle fatigue regime. On the other hand, in the high cycle fatigue regime, the specimens deep rolled at a temperature of 200 °C show slightly lower fatigue lifetimes as compared to the room-temperature deep rolled as-quenched condition (see Fig. 10). This can be attributed to the lower near-surface macroscopic compressive residual stresses as well as work hardening states after deep rolling at a temperature of 200 °C (see Fig. 8).

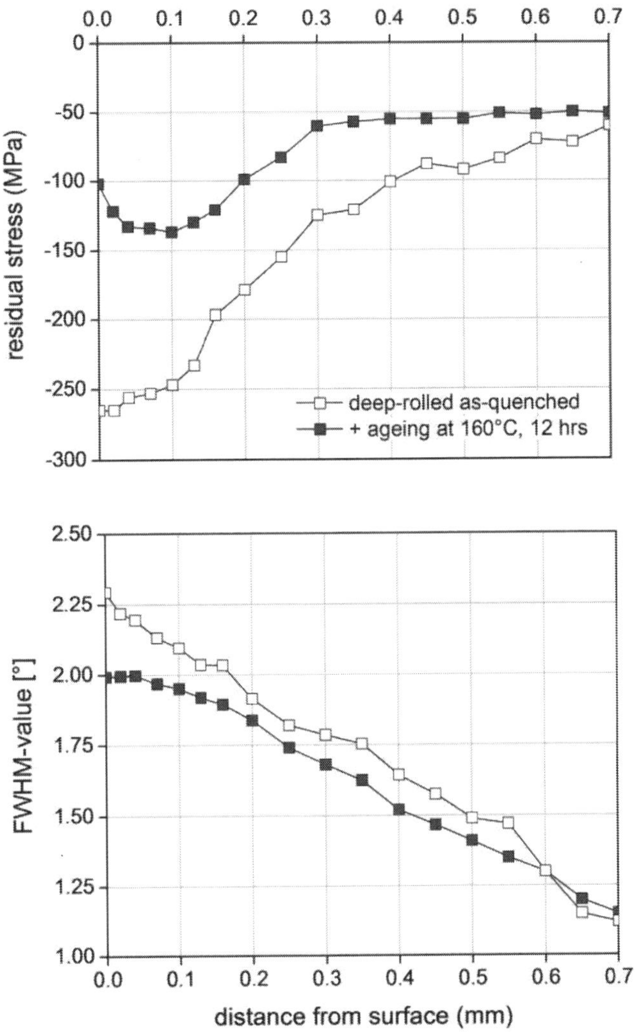

Figure 4. Residual stress- and FWHM-depth-profiles of deep rolled as-quenched AA6110 before and after the optimized ageing treatment.

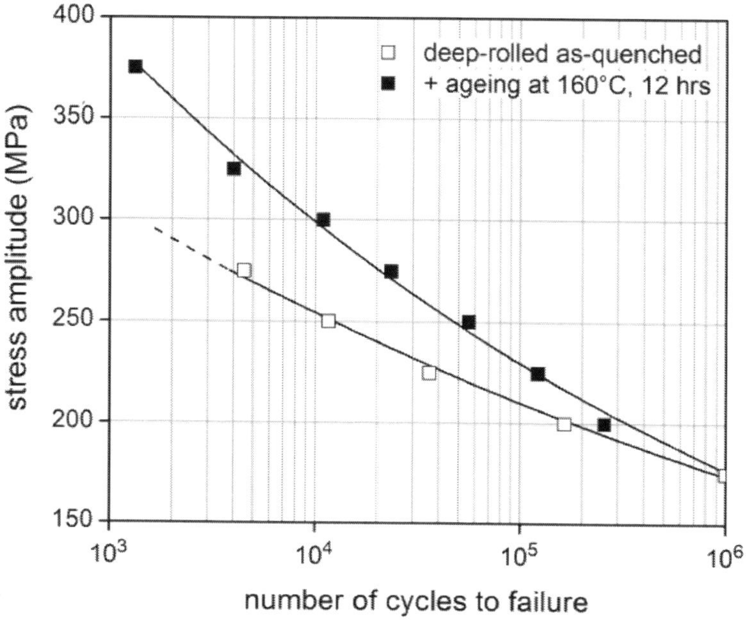

Figure 5. Non-statistically evaluated s/n curves of deep rolled as-quenched specimens before and after the optimized ageing treatment.

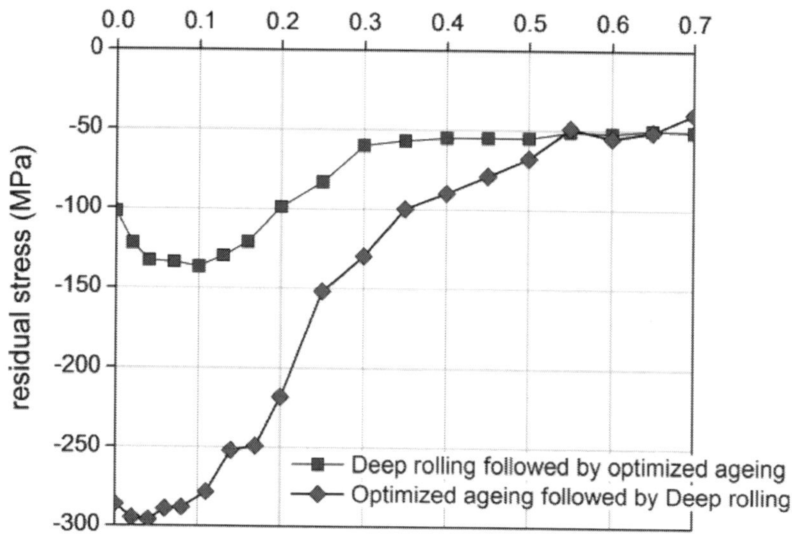

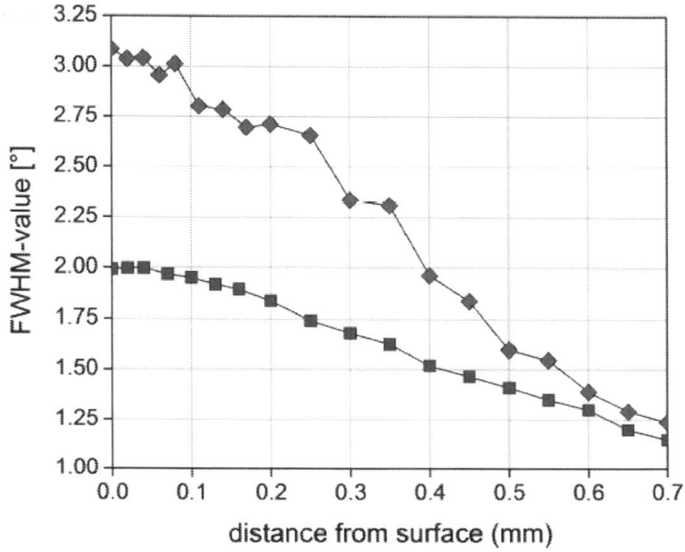

Figure 6. Residual stress- and FWHM-depth-profiles of deep rolled as-quenched AA6110 after the optimized ageing treatment as compared to deep rolled optimized-aged AA6110.

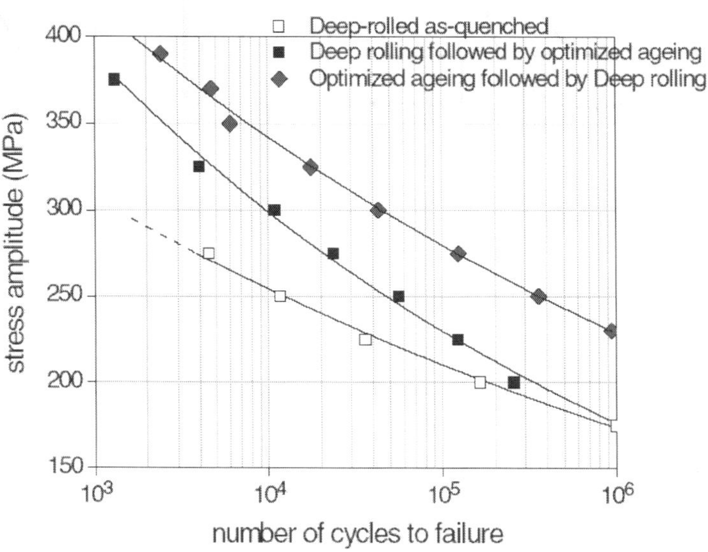

Figure 7. Non-statistically evaluated s/n curves of optimized-aged deep rolled as-quenched AA6110 as compared to deep rolled optimized-aged AA6110 as well as deep rolled as-quenched AA6110.

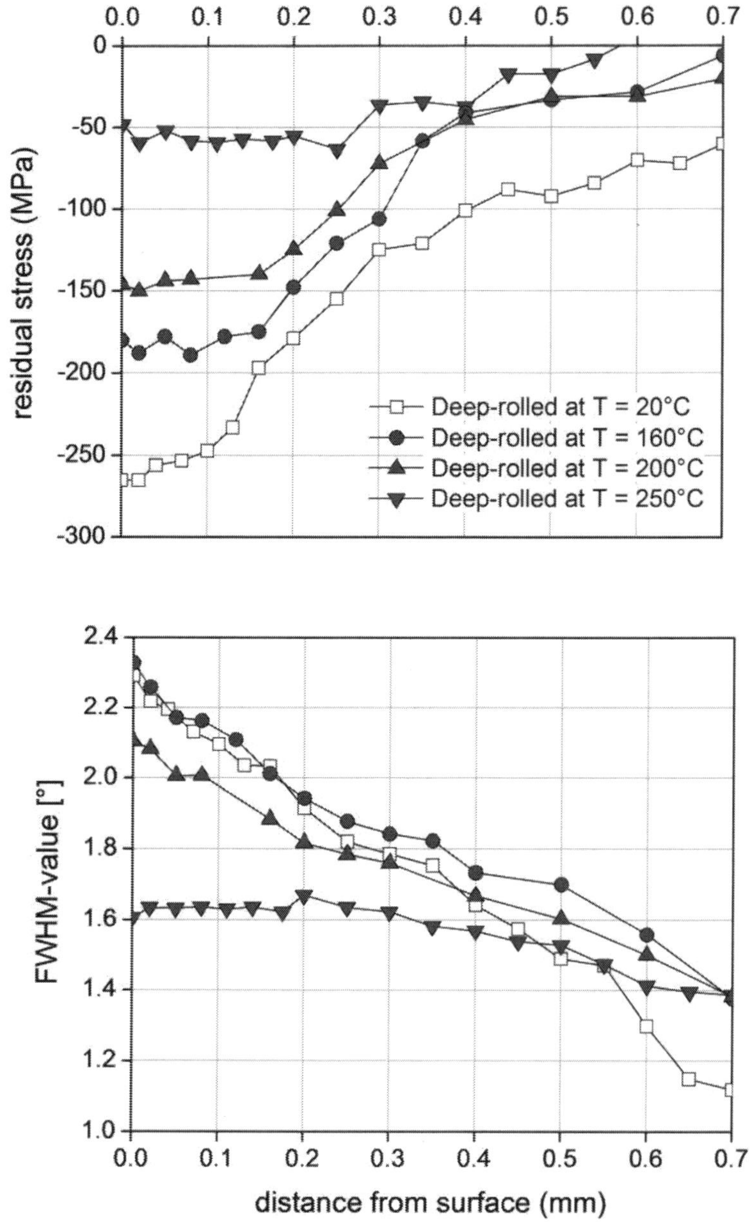

Figure 8. Depth-profiles of near-surface macroscopic compressive residual stresses and FWHM-values of high-temperature deep rolled as-quenched AA6110 for different deep rolling temperatures.

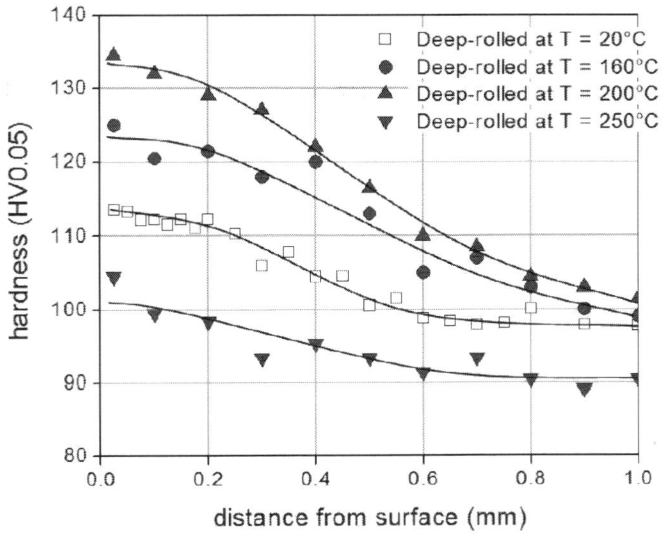

Figure 9. Depth-profiles of near-surface hardness values of high-temperature deep rolled as-quenched AA6110 for different deep rolling temperatures.

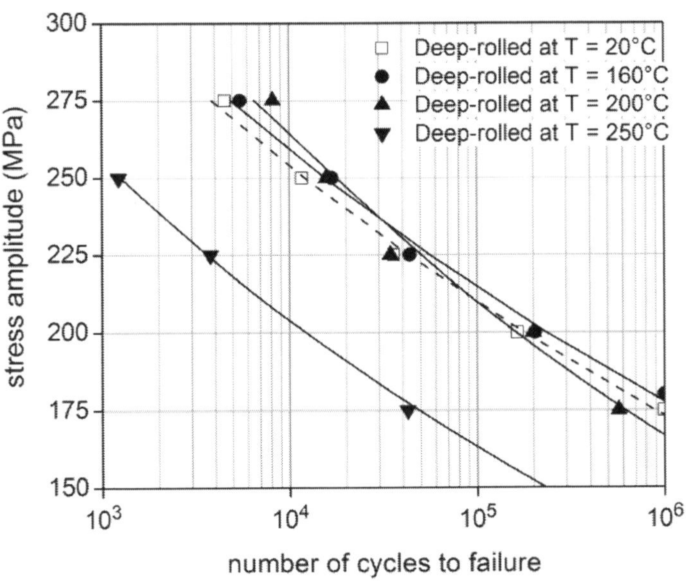

Figure 10. Non-statistically evaluated s/n-curves of high-temperature deep rolled as-quenched AA6110 for different deep rolling temperatures.

4. CONCLUSIONS

1. Deep rolling combined with an ageing treatment of as-quenched AA6110 resulted in an increase of near-surface hardness values. At the same time, a reduction of near-surface macroscopic compressive residual stresses and work hardening states takes place. The optimized ageing condition is an ageing temperature of 160 °C for about 12 hr.

2. Increased near-surface hardness values after the optimized ageing treatment seem to be very beneficial for the fatigue lifetimes in low cycle fatigue regime, whereas in the high cycle fatigue regime, macroscopic compressive residual stresses are additionally important.

3. Deep rolling at elevated temperatures up to approximately 200 °C resulted in an increase of near-surface hardness values in the as-quenched condition, whereas lower macroscopic compressive residual stresses and work hardening states were observed because static/dynamic recovery processes occurred. Deep rolling at a temperature of 250 °C exhibits detrimental effects on near-surface properties due to the over-ageing effects and high-rate static/dynamic recovery at this temperature.

4. An optimized ageing treatment followed by a conventional deep rolling treatment, can completely combine precipitation hardening and work hardening in near-surface regions of AA6110 into an optimized microstructure and residual stress state and thus result in the best surface properties and fatigue lifetime of the investigated AA6110.

5. ACKNOWLEDGEMENTS

The authors would like to express sincere thanks to the German Science Foundation (DFG) and to the Faculty of Engineering, Kasetsart University, Thailand, for financial support for Dr.-Ing. I. Altenberger and Dr.-Ing. P. Juijerm, respectively.

REFERENCES

1. J. Polmear, 1996 Light Alloys: Metallurgy of the Light Metals. London Halsted Press.
2. ASM 1993 ASM Metal Handbook: 2 Properties and Selection: Nonferrous Alloys and Special-Purpose Materials. Ohio ASM International
3. Niku-Lari, Ed. 1987 Advances in Surface Treatments. Oxford Pergamon Press
4. Scholtes, 1997 Assessment of residual stresses. in: V. Hauk (Ed.). Structural and Residual Stress Analysis by Nondestructive Methods. Amsterdam Elsevier
5. Altenberger, R. K. Nalla, Y. Sano, L. Wagner, R. O. Ritchie, 2012 On the effect of deep-rolling and laser-peening on the stress-controlled low- and high-cycle fatigue behavior of Ti-6Al-4V at elevated temperatures up to 550 °C. Int. J. Fatigue; in press.
6. Altenberger, B. Scholtes, 1999 Improvement of fatigue behaviour of mechanically surface treated materials by annealing. Scrip Mater 41 8 873 881
7. R. Menig, V. Schulze, O. Vöhringer, 2002 Effects of static aging on residual stress stability and alternating bending strength of shot peened AISI 4140. Z Metallkd 93 7 635 640
8. P. Juijerm, U. Noster, I. Altenberger, B. Scholtes, 2004 Mater. Sci. Eng. A379 286
9. P. Juijerm, I. Altenberger, B. Scholtes, 2006 Mater. Sci. Eng A 426 4
10. P. Juijerm, I. Altenberger, 2006 Scrip. Mater 55 943
11. P. Juijerm, I. Altenberger, 2006 Scrip. Mater 55 1111
12. K. Gregory, C. Müller, L. Wagner, 1993 Bevorzugte Randschichtaushärtung: Neue Verfahren zur Verbesserung des Dauerschwingverhaltens mecha- nisch belasteter Bauteile, Metall 47 915
13. Wagner, A. Berg, T. Dörr, M. Hilpert, 2000 Kugelstrahlen und Festwalzen von Titan-, Auminium-, und Magnesiumlegierungen, in: H. Wohlfahrt and P. Krull (Eds.), Mechanische Oberflächenbehandlungen Wiley-VCH Weinheim

14. Berg, J. Kiese, L. Wagner, 1998 Microstructural gradients in Ti-3Al-86Cr-4Zr -4Mo for excellent HCF strength and toughness Mater. Sci. Eng. A 243 146 149

15. K. Gregory, L. Wagner, 2003 Property improvement in light metals using shot peening, in: L. Wagner (Ed.), Shot Peening, Wiley-VCH, Weinheim 349

16. F. J. Hamphreys, M. Haltherly, 1995 Recrystallization and Related Annealing Phenomena, Pergamon Oxford

CHAPTER 7

Phase Evolution and Mechanical Behavior of the Semi-Solid SIMA Processed 7075 Aluminum Alloy

Behzad Binesh and Mehrdad Aghaie-Khafri *

Faculty of Materials Science and Engineering,
K. N. Toosi University of Technology, Tehran 1999143344, Iran

ABSTRACT

Microstructural and mechanical behaviors of semi-solid 7075 aluminum alloy were investigated during semi-solid processing. The strain induced melt activation (SIMA) process consisted of applying uniaxial compression strain at ambient temperature and subsequent semi-solid treatment at 600–620 °C for 5–35 min. Microstructures were characterized by scanning electron microscope (SEM), energy dispersive spectroscopy (EDS), and X-ray diffraction (XRD). During the isothermal heating, intermetallic precipitates were gradually dissolved through the phase transformations of α-Al + η ($MgZn_2$) → liquid phase (L) and then α-Al + Al_2CuMg (S) + Mg_2Si → liquid phase (L). However, Fe-rich precipitates appeared mainly as square particles at the grain boundaries at low heating temperatures. Cu and Si were enriched at the grain boundaries during the isothermal treatment while a significant depletion of Mg was also observed at the grain boundaries. The mechanical behavior of different SIMA processed samples in the semi-solid state were investigated by means of hot compression tests. The results indicated that the SIMA processed sample with near equiaxed microstructure exhibits the highest flow resistance during thixoforming which significantly decreases in the case of samples with globular microstructures. This was justified based on the governing deformation mechanisms for different thixoformed microstructures.

KEYWORDS

7075 Al alloy; SIMA; phase transformation; semi-solid; thixoforming

1. INTRODUCTION

Semi-solid metal processing is a single step fabrication method for production of near net shape metallic parts [1,2,3]. Advantages of semi-solid processing when it is compared with the conventional casting, forging, and powder metallurgy processes are (i) reduction of the porosity, macrosegregation, and surface cracks; (ii) high dimensional accuracy of the produced parts; (iii) extending of the forming die life due to decreasing of the casting temperature and required forming force; and (iv) the uniform filling of the die and reduction of the manufacturing stages and costs [4,5].

The main goal in the semi-solid forming processes is to achieve a thixotropic microstructure consisting of fine and spherical solid grains which are uniformly distributed in the liquid matrix [6,7]. Various routes have been developed for production of thixotropic aluminum, magnesium, lead, and also ferrous alloys including mechanical and magneto-hydrodynamic (MHD) stirring processes, the thermo-mechanical strain induced melt activation (SIMA) process and the spray casting process [7,8,9]. The SIMA process which is based on the mechanical work was introduced by Young *et al.* [10] in the 1980s. This process has received great attention in recent years due to its simplicity, low equipment costs, and its capability to process many engineering alloys [8]. The two main stages of the SIMA process are cold or warm plastic deformation and then partial remelting in the semi-solid range [11,12].

Many studies have been focused on the semi-solid microstructure of cast aluminum alloys due to superior semi-solid formability [12,13]. However, the cast products show poorer mechanical properties than wrought alloys e.g., 2xxx, 6xxx, and especially 7xxx series. This restricts their applicability in the high strength applications such as aerospace and automotive. On the other hand, control of the semi-solid heat treatment in the wrought alloys is relatively difficult due to their wide solidification range and high sensitivity of solid fraction to temperature variations (steep slope of the solid fraction *versus*temperature). Furthermore, the wrought alloys are severely susceptible to formation of hot tearing during the forming process [14]. Consequently, investigation on the semi-solid microstructure of wrought aluminum alloys to achieve an appropriate thixotropic microstructure is vital. Yong *et al.* [15] prepared an appropriate 7075 semi-solid feedstock by 50% swaging of the specimens at room temperature and then heat treating at 590 °C. They also investigated the properties of the thixoforged part. Multi step heat treatment in thixoforming processes was investigated by Chayong *et al.* [16,17]. Atkinson *et al.* [18] studied the recrystallization of 7075 Al alloy during the recrystallization and partial melting (RAP) process. Vaneetveld *et al.* [19] investigated the effects of heating parameters on the recrystallization in the RAP process and suggested using the high solid fraction to prevent hot tearing in the thixoforging process. Bolouri *et al.* [20] studied the effect of cold work on the semi-solid microstructure by compression straining of the samples up to 40% and Mohammadi *et al.* [21] investigated the effects of the isothermal treatment on the semi-solid microstructure of extruded samples.

Previous studies on the semi-solid processing of 7075 aluminum alloy were mainly restricted to investigations in the field of thixotropic feedstock preparation criteria and producing some thixoformed parts. However, detailed information of the phase transformations during partial remelting and the correlation between the initial microstructure and rheological behavior as well as their final mechanical properties needs to be further investigated. There is a lack of thorough research on this topic in the literature. Consequently, the present work aims to investigate the phase evolution and kinetics of partial remelting during isothermal treatment in the mushy zone and the semi-solid deformation behavior of AA7075 alloy.

2. EXPERIMENTAL PROCEDURE

2.1. Materials and Thermal Analysis

Commercial extruded round bar of 7075-T6 aluminum alloy was used as the starting material, the chemical composition of which is shown in Table 1.

Table 1. Chemical composition of wrought 7075 Al alloy in wt. %.

Element	Al	Mn	Fe	Cr	Cu	Mg	Zn	Si
wt. %	Bal.	0.28	0.28	0.13	1.58	2.41	5.31	0.14

In order to determine the solidus and liquidus temperatures and the solid volume fraction *versus* the temperature curve of 7075 alloy, thermal analysis was carried out using a Metler Star SW 10 differential scanning calorimeter (DSC) apparatus, Metler, Giessen, Germany. Samples of the material (30 mg) were put into the alumina pan and then heated to 700 °C at 10 °C/min under nitrogen atmosphere. The solid volume fraction-temperature relationship was determined using the obtained DSC curve.

2.2. SIMA Process

The SIMA process was carried out to obtain a thixotropic microstructure in 7075 Al alloy. Cylindrical samples with a diameter of 30 mm and height of 35 mm were machined from the starting material. The samples were stress relieved at 460 °C for 1 h, and then air cooled. Compression of cylindrical samples, with a compression ratio of 20%, was carried out at room temperature using a high strain rate (~15 s^{-1}) hammer apparatus. The compressed specimens were machined to samples with a diameter of 20 mm and height of 15 mm to ensure uniform strain in the core section. Isothermal treatment of samples was carried out at three different temperatures of 600, 610, and 620 °C with an accuracy of ±1 °C for 5–35 min in a resistance furnace. Heating cycles were interrupted at predetermined interval times and then samples were quenched to study the microstructure.

2.3. Compression Tests

The thixoforming process consisting of hot compression experiments at the semi-solid temperature range was carried out on the specimens of different initial microstructures. The cylindrical specimens with 11 mm in height and 7 mm in diameter, according to ASTM E209 standard, were prepared from the SIMA processed samples. The isothermal hot compression tests were carried out at temperatures of 600 and 610 °C and strain rate of 0.3 s^{-1} using a Gotech-Al7000 model universal testing machine, Gotech, Taichung, Taiwan, equipped with an electrical resistance furnace. The specimens were first preheated to the preset temperature and soaked for 5 min to homogenize the samples. The specimens were then compressed up to a strain of 0.6 followed by rapid quenching in water. A high accuracy load cell was used to record the true stress values, and the true strain values were computed using the displacement data. A very thin mica plate was used to reduce the friction effect and to prevent the adhesion of the specimen on the die. The specimens were sectioned parallel to the deformation direction to study the microstructural changes.

2.4. Metallography and Microstructural Characterization

The microstructure of the quenched samples was examined using the standard metallography method. The surface perpendicular to the direction of the compression was ground with standard SiC abrasive papers and polished with 0.25 μm diamond paste. Samples were etched using modified Keller etchant solution (3 mL HF, 2 mL HCl; 20 mL HNO_3; 175 mL H_2O). Microstructural investigations were carried out using a Neophot 32 optical microscope, Norfab, Trondheim, Norway, and Vega©Tescan, Tescan, Libušina, Czech Republic, and Mira3Tescan field emission scanning electron microscope, Tescan, Libušina, Czech Republic, equipped with EDS detector (Tescan, Libušina, Czech Republic). The linear intercept method was used to determine the average grain size. A series of straight lines with a specified length was considered on the optical micrographs for each sample and the average grain size was determined by division of the line length (L) by the number of intercepted grains (N). The shape factor of the solid grains was measured by means of Clemex professional edition image analyzer software (version 5.0, Clemex Technologies Inc., Longueuil, QC, Canada) and using Equation (1) [22]:

$$F = \frac{\sum_{N=1}^{N} 4\pi A/P^2}{N} \tag{1}$$

where A and P are area and perimeter of the solid grains, respectively, and N is the number of grains. For each sample, measurements were taken from the whole sectioned area including 300–400 solid particles per sample.

X-ray diffraction (XRD) analysis was used to investigate phase evolution in the semi-solid treated samples. XRD experiment was carried out using a Philips X'pert model apparatus, Panalytical, Eindhoven, The Netherlands, with CuKα target and wave length of 0.15406 nm.

3. RESULTS AND DISCUSSION

3.1. Microstructure and Thermal Analysis of the Starting 7075 Alloy

Figure 1 shows the scanning electron microscope (SEM) micrographs of the as-received 7075 aluminum alloy. It can be observed (Figure 1a,b) that the microstructure of the alloy consists of directed α-Al grains and some precipitate particles in the direction of the initial extrusion process. It is worth noting that Zn, Cu, and Mg are the main alloying elements in 7075 alloy which play an important role in the formation of precipitates. In the 7xxx series of aluminum alloys, when the Zn:Mg ratios are between 1:2 and 1:3, $MgZn_2$ (η) precipitates are produced at ageing temperatures below 200 °C and are the main strengthening factor in 7xxx alloys [18]. The dimensions of these precipitates are only a few tens of nanometer which can only be revealed by TEM technique. Considering Figure 1b–d, EDS analysis results of the coarse precipitate particles, as summarized in Table 2, suggested them to be (I) Al_2CuMg intermetallic phase (which is usually denoted as S-phase) with relatively smooth corners (C_1 and C_2 in Figure 1b,d), (II) $Al_6(Cu,Fe)$ intermetallic phase which is also known as $Al_{23}Fe_4Cu$ or α(FeCu) with polygonal (B_1 in Figure 1b) or irregular (B_2 in Figure 1c) shapes, (III) Al_7Cu_2Fe (β) phase with irregular shapes which have been elongated along the extrusion direction (A in Figure 1b) and (IV) dark Mg_2Si precipitates with irregular shapes (D in Figure 1c). Considering the EDS analysis results, Fe-rich intermetallic phases ($Al_6(Cu,Fe)$ and Al_7Cu_2Fe) also contain small amounts of manganese which can be substituted for iron as a result of their close atomic radii.

Figure 2a shows the DSC curve of the 7075 aluminum sample. The solidus and liquidus temperatures based on the DSC curve were 486 and 649 °C, respectively. The characteristic values of the areas under the DSC peaks calculated for mass unity are also included in Figure 2a. The liquid or solid fraction *versus* the temperature curve was obtained by integration of the DSC curve (Figure 2b). The results were used to determine the isothermal heat treatment temperatures in the SIMA process. It is worth noting that the liquid or solid volume fraction at a known temperature based on Figure 2b is not exactly the same as its values appearing in the micrographs. This is due to insufficient cooling rate of the semi-solid samples after isothermal treatment. In other words, some of the liquid fraction that transforms to the solid during quenching is considered as solid fraction and therefore the liquid volume fraction observed in the micrographs is always lower than that determined by Figure 2b.

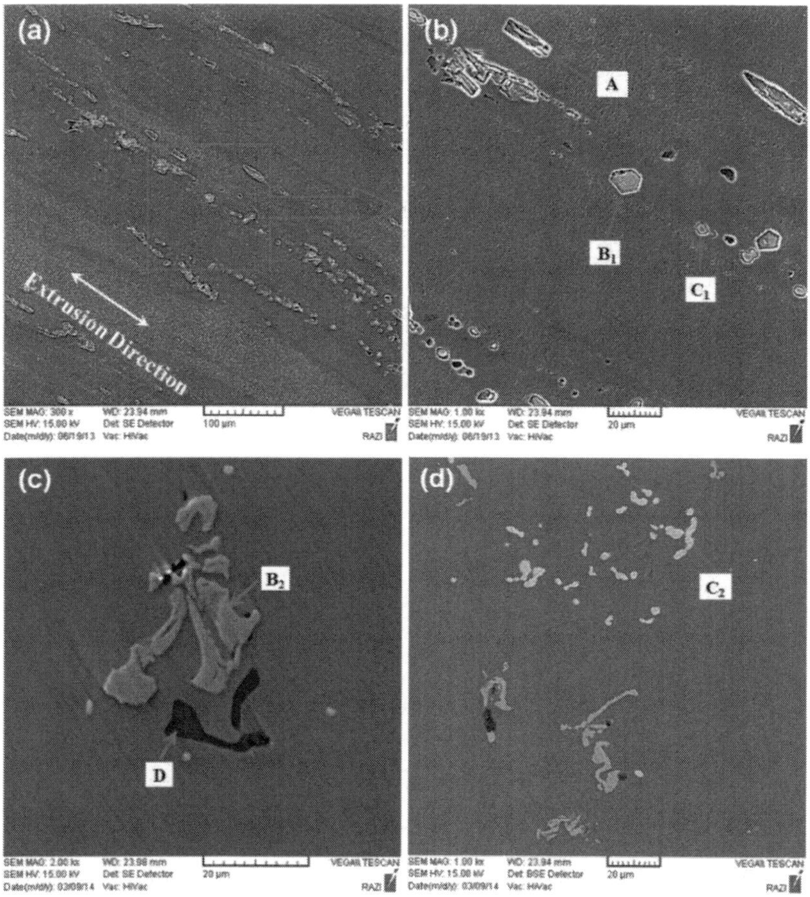

Figure 1. (**a**) Scanning electron microscope (SEM) secondary electron image; (**b**) SEM secondary electron image of selected area in (a); (**c**) SEM secondary electron image; (**d**) SEM backscattered electron image of as-received 7075 sample.

3.2. Microstructural Evolution during Partial Remelting of the Compressed 7075 Alloy

Microstructures of 7075 specimens after 20% compression and isothermal heat treating at the semi-solid temperature of 600 °C for various times during the SIMA process are shown in Figure 3. The initial microstructure consisted of plastically deformed grains that gradually transforms to a globular microstructure of spherical and equiaxed grains after heating and partial remelting at 600 °C. Micrographs of Figure 3a,b show that following 5 min heating of the sample at 600 °C, no evidence of recrystallization was observed. Following the increase of the heating time up to 10 min, a near equiaxed microstructure consisting of elongated and polygonal solid grains with what appears to be quenched liquid at the grain boundaries was formed (Figure 3c and Figure 4a). In some zones, the size of these grains is quite small which implies a recrystallization phenomenon. Moreover, the formation of the liquid phase is expected to be as a result of partial remelting during isothermal holding at 600 °C. The results of research conducted by Atkinson *et al.* [18] revealed that liquids are first formed at the recrystallized grain boundaries of RAP processed 7075 alloy because these are areas with higher solute concentrations. However, the liquid phase has not penetrated thoroughly into the grain boundaries which results in incomplete grain boundary wetting, as shown in Figure 4a. The mechanism of low melting point phase formation in the semi-solid microstructure is extensively discussed in Section 3.4. With further extension of the heating time up to 15 min, the whole microstructure was transformed into individual polygonal grains due to separation and increase of the liquid phase (Figure 3d). The grain size significantly increased compared with that of the grains which had just recrystallized and separated, shown in Figure 3c. This indicates that the separation and coarsening were in competition.

Table 2. Scanning electron microscope (SEM)/energy dispersive spectroscopy (EDS) analysis results of selected constituent particles highlighted in Figure 1.

Content of Elements	Constituent Particles (at. %)					
	A	B_1	B_2	C_1	C_2	D
Al	61.35	78.39	82.23	66.72	57.59	-
Zn	0.31	0.20	1.19	0.17	0.05	-
Cu	20.00	5.30	2.67	17.16	18.64	-
Mg	0.03	0.02	0.81	13.83	23.38	69.42
Fe	13.71	11.61	10.64	0.52	0.02	-
Si	1.79	0.13	1.09	0.47	0.03	30.58
Mn	2.27	3.56	1.33	0.52	0.18	-
Cr	0.55	0.79	0.04	0.59	0.10	-
Phase shape	Irregular	polygonal	Irregular	Spherical	Irregular with round corners	Dark-Irregular
Closest phase	Al_7Cu_2Fe	$Al_6(Cu,Fe)$	$Al_6(Cu,Fe)$	Al_2CuMg	Al_2CuMg	Mg_2Si

Comparing the primary grain size of the deformed sample and the resulting polygonal particle size, it can be suggested that the recrystallization and separation dominated during the period of 5 min to

10 min or at a longer time and then the coarsening became the dominant process. During the coarsening stage, it was found that the liquid amount is relatively small and the neighboring grains are only separated by a thin liquid layer (Figure 3c,d).

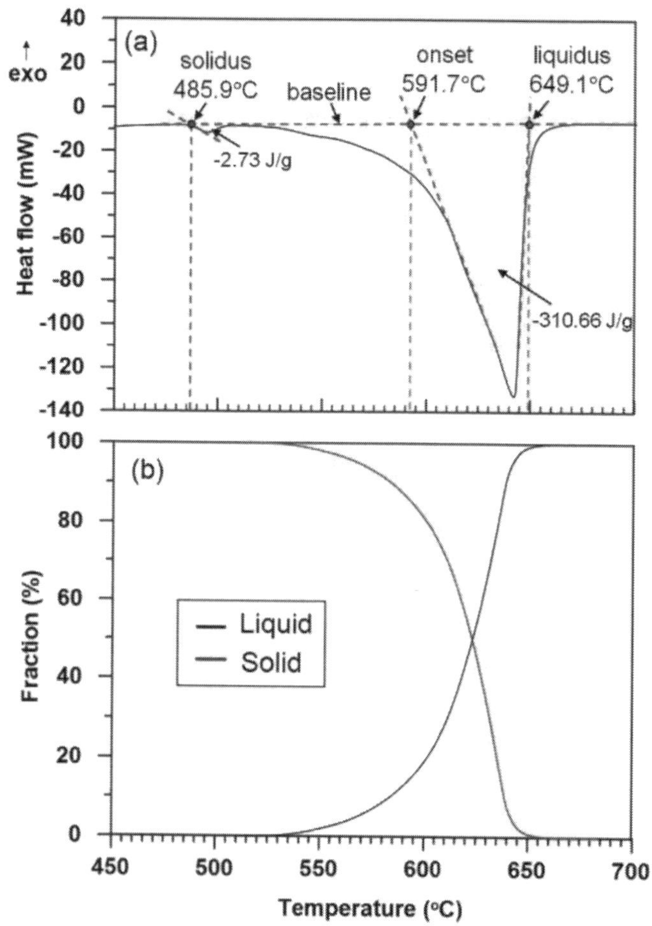

Figure 2. (**a**) Differential scanning calorimeter (DSC) curve of 7075 Al alloy at a heating rate of 10 °C/min; (**b**) solid and liquid volume fraction *versus* temperature derived from the DSC curve.

During the isothermal heating period of 15 min to 25 min, the interconnection between the solid grains reduced and the grains gradually became spherical (Figure 3e). The spheroidization mainly occurred during this stage. During the period of 25 min to 35 min, the liquid phase obviously increased (comparing Figure 3e,f) which finally led to intergranular formation of thick eutectic regions following 35 min heating of the sample (indicated by the E letter in Figure 3f). This is evidenced with the eutectic phase becoming the predominant feature at grain boundaries after semi-solid soaking (Figure 4b). It can be expected that the coarsening of spherical grains will occur with raising of the heating time over 35 min.

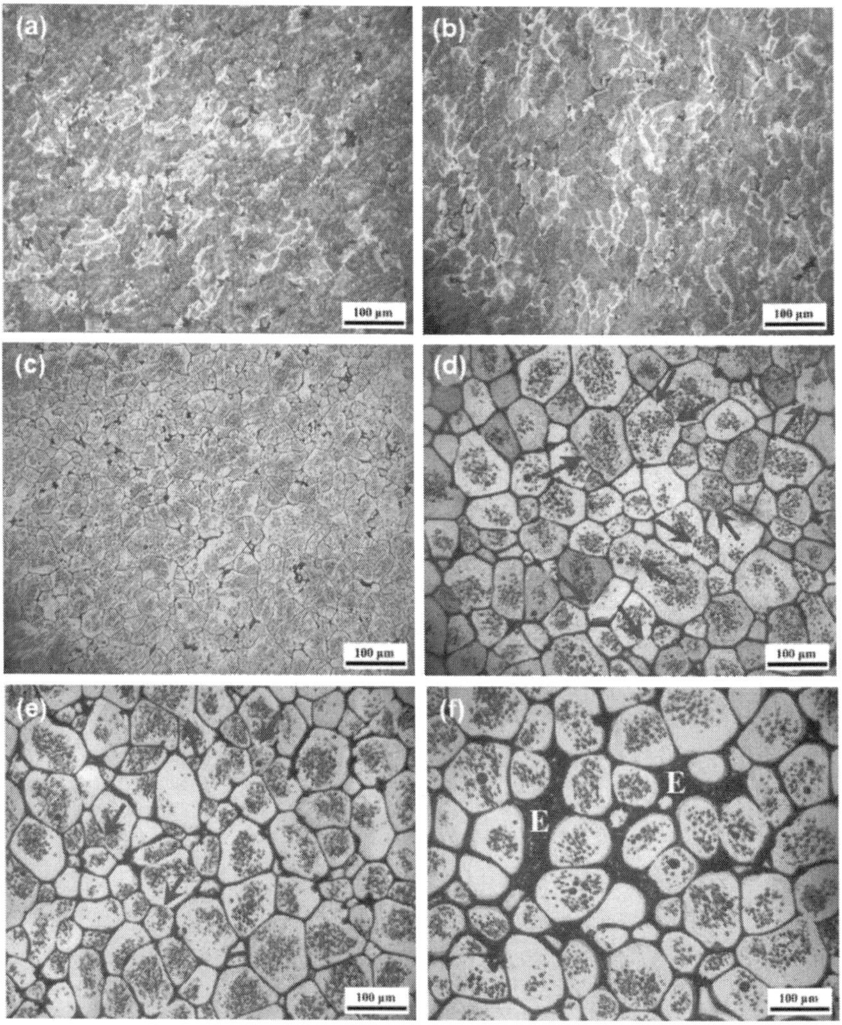

Figure 3. Optical micrographs of samples compressed 20% and isothermally heat treated at 600 °C for (**a**) 0 min; (**b**) 5 min; (**c**) 10 min; (**d**) 15 min; (**e**) 25 min; (**f**) 35 min.

In addition, some intragranular islands which are claimed to be liquid pools just before quenching, were observed within the solid grains.

Therefore, the semi-solid microstructure consisted of α-Al spherical solid grains, intergranular eutectic liquid film and entrapped liquid droplets within the solid grains. Entrapped liquid droplets were produced in two different ways. The first was as a consequence of alloying elements segregation within the solid particles during the partial remelting. In this case, some fine liquid droplets entrap within the solid grains. The second one was due to coalescence of solid grains with complex geometrical shape during the heating stage to decrease the solid-liquid phase interfacial energy. In this condition, larger droplets of the liquid phase, in comparison with the former, were created within the grains [23]. As is observed in Figure 3, with longer holding times some of the fine liquid droplets as a result of coalescence phenomena or atomic diffusion joined to the larger liquid droplets. Subsequently, these new liquid droplets become more spherical to reduce the solid/liquid interfacial energy. Finally, with further increase of the holding time, a lesser number of the entrapped liquid droplets was produced owing to diffusion into the liquid phase matrix that surrounded the solid grains.

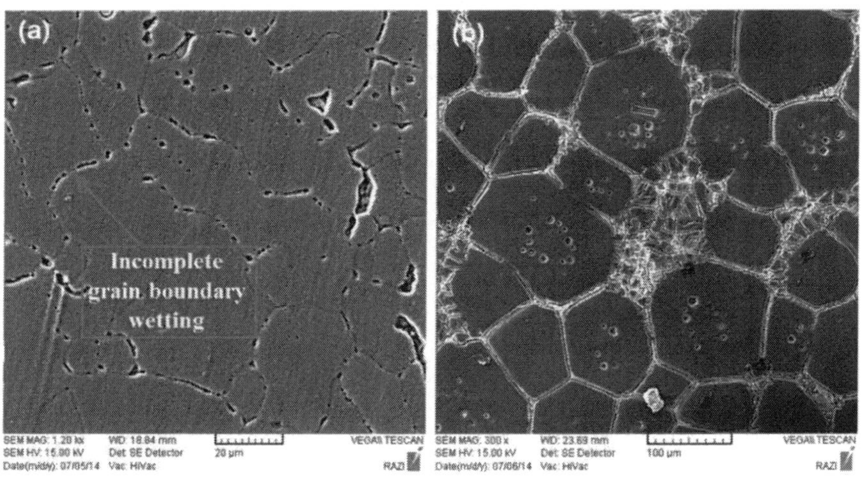

Figure 4. SEM images of 20% compressed 7075 alloy heat treated at 600 °C for (**a**) 10 min; (**b**) 35 min.

3.3. Kinetics of the Microstructural Evolution during Partial Remelting

Considering the obtained results in the present study, the kinetics of the microstructural evolution of 7075 samples during the SIMA process can be divided into four steps. (I) recovery, recrystallization and structural separation, (II) coarsening of polygonal solid grains, (III) spheroidization of polygonal solid grains and (IV) coarsening of spherical grains. Recovery, recrystallization, and structural separation mainly occur in the early stages of the isothermal heating operation. During recovery and recrystallization, vacancies are joined to each other and dislocations are rearranged as low energy structures by climbing or cross slipping and subgrain boundaries are created. In this step, grains of high dislocation density are replaced by new subgrains with less dislocation density. Moreover, because the holding temperature is higher than the eutectic line, partial remelting also occurred [20,23]. It has been found that when the angle between subgrains is greater than 20°, the surface energy of the grain boundaries is greater than two fold of the solid/liquid interfacial energy ($\gamma_{gb} > 2\gamma_{sl}$). If these grain boundaries contact with liquid phase, the grain boundaries are replaced by a thin liquid film. Considering the*vice versa* case *i.e.* $\gamma_{gb} < 2\gamma_{sl}$, low energy grain boundaries will be formed which are not wetted by the liquid phase. If two solid grains reach together through such a low energy boundary, they can be joined together and stimulate the coalescence of solid particles [24].

With rising of the isothermal heating time, the liquid volume fraction becomes greater and isolated polygonal solid grains are created. By continuing the heating process, these grains grow and coarsen. The growth and coarsening of the solid grains in the SIMA process are controlled by two mechanisms of coalescence and Ostwald ripening. During the growing and coarsening of the microstructure, diffusion of the solid material from regions with high curvature to low curvature points occurs based on Equation (2) [12]. This subsequently provides the required driving force for spheroidization of solid grains.

$$\Delta T_r = -\frac{2\sigma T_m V_s k}{\Delta H} \tag{2}$$

where $\Delta T_r = T_m - T$ represents the difference to the equilibrium melting point, T_m is the equilibrium transformation temperature, k is the mean surface curvature of the solid particles, σ is the surface tension and ΔH (a negative value) is the molar change in the enthalpies of the solid and liquid.

As long as the grains are not spherical during semi-solid treatment, the spheroidization process leads to change of the particles shape from polygonal to spherical where the numbers per unit volume remain constant. Following this stage, smaller grains which have a lower melting point according to Equation (2), are melted in favor of the larger grains and the numbers per unit volume are reduced. The microstructure of samples that have been subjected to partial remelting in the SIMA process, are evolved by one of the mentioned mechanisms depending on the sphericity of solid grains and the amount of the liquid volume fraction [25].

Ostwald ripening is a diffusion controlled mechanism and acts as the dominant mechanism at high liquid volume fractions *i.e.* high heating temperatures and long holding times. This mechanism is less effective in particle growth but has a great influence on the spheroidization of solid grains. In contrast, the coalescence mechanism which needs shorter holding times and a small amount of liquid fractions, is more effective in grain growth and has a minor effect on the spheroidization process [25,26,27].

When the liquid fraction is low (early stages of heating operation), the solid grains are readily in contact with one another, and coalescence is the dominant mechanism in the coarsening of the microstructure. This can be verified by considering the microscopic images shown in Figure 3d–f. The arrows on the micrographs show that some grain boundaries are removed due to coalescence of the solid grains. Therefore, it can be concluded that the growing and coarsening of grains in the sample heated at 600 °C for 15 min mainly occurs through the coalescence mechanism due to the relatively low liquid fraction (Figure 3d). With further increase of the holding time (Figure 3e,f), the liquid fraction increased and Ostwald ripening mechanism was more effective in the coarsening process. When the majority of the solid grains gained uniform surface curvature through these two mechanisms, the growth of spherical grains commences. Therefore, spheroidization and further growth of solid grains by the two mentioned competing mechanisms are expected in the SIMA process [25].

3.4. Phase Evolution and Alloying Element Distribution during Partial Remelting of Compressed 7075 Alloy

A close examination of the micrographs in Figure 3 reveals that the volume fractions of precipitates in the microstructure reduce with the rise of the isothermal holding time. This is clearly confirmed by increasing the brightness of the recrystallized solid grains. XRD results shown in Figure 5 also confirm this phenomenon. The X-ray diffraction pattern of the 20% compressed sample prior to the isothermal treatment, indicated that the microstructure consists of α-Al phase and η, S, $Al_6(Cu,Fe)$, Al_7Cu_2Fe, and Mg_2Si intermetallic phases, which is similar to the as-received sample (Figure 1). According to the X-ray diffraction patterns of samples heat treated at 600 °C for 5 to 35 min, it is clear that the intensity of η, S, and

Mg_2Si peaks decrease with the increase of the holding time. The diffraction peaks of the 20% compressed sample are shown by symbols and the corresponding peaks in each case are specified by arrows for the other samples in Figure 5. The diffraction peaks of η, S, and Mg_2Si phases almost disappeared after heating for 25 min. However, some diffraction peaks of Fe-rich intermetallic phases (Al_7Cu_2Fe and $Al_6(Cu, Fe)$) show little change with prolonging of the holding time. Considering the sample heated for 35 min, the presence of Al_7Cu_2Fe and $Al_6(Cu, Fe)$ phases was distinguishable. However, no evidence of η, S, and Mg_2Si phases was observed. This illustrates that Fe-rich secondary particles cannot be dissolved into the matrix for the semi-solid treatment conditions used in the present study.

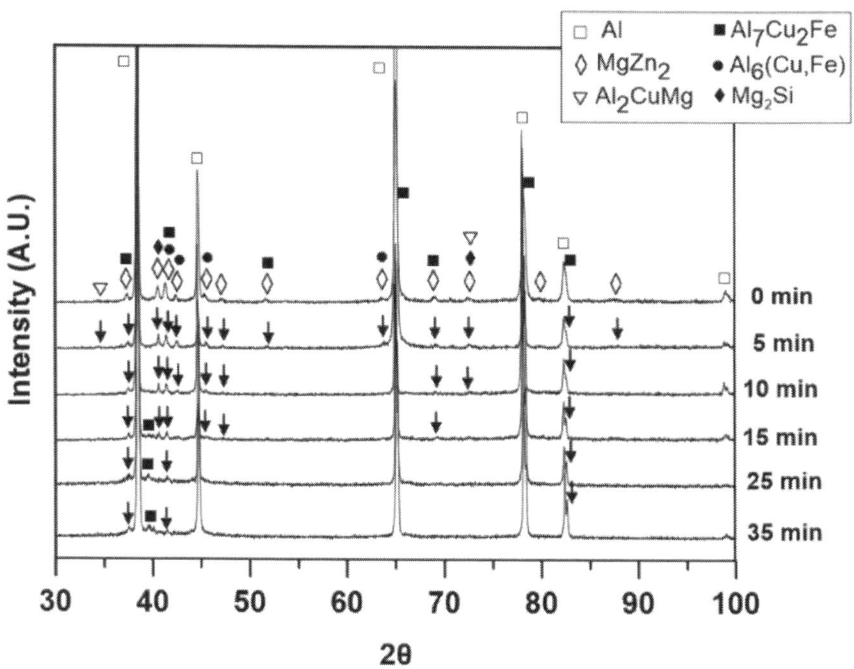

Figure 5. X-ray diffraction (XRD) patterns of 20% compressed 7075 alloys, heat treated at 600 °C for different times and then water quenched.

Previous work reported that the melting temperature of η-phase in 7xxx alloys is 475–480 °C [26,27]. This is fairly well correlated with the first peak in the DSC curve of the starting material with an enthalpy of 2.7 J/g in Figure 1. The solution temperatures of S and Mg_2Si phases are 490–501 °C [28,29] and 478–525 °C [30], respectively, depending on the Mg content of the alloy. Considering the XRD results (Figure 5) and solution temperature of intermetallic precipitates, it can be suggested that the dissolving of intermetallic particles commences with melting of η precipitates after heating for 10 min through the reaction of α-Al + η → L. Following the increase of the heating time, dissolving of S and Mg_2Si intermetallic particles also occurs through the phase transformation of α-Al + S + Mg_2Si → L. These precipitates are fully dissolved after isothermal heating for 25 min.

The Fe-rich intermetallic precipitates have higher solution temperatures and are usually stable over 600 °C [31]. Microstructural observations revealed that the Fe-rich intermetallic particles, shown in Figure 6, precipitate at grain boundaries during partial remelting. Precipitation of these particles at the grain boundaries has a great influence on pinning of the grain boundaries movement. Therefore, it may cause the retardation of the particle growth and coarsening during the isothermal treatment. This is discussed in more detail in Section 3.5.

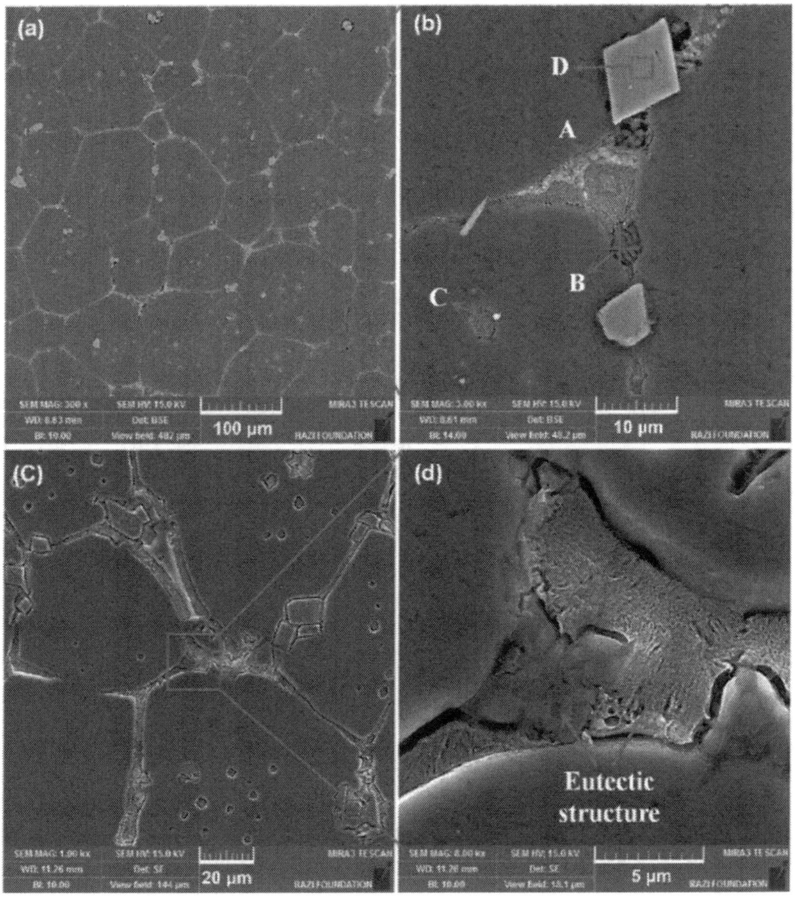

Figure 6. SEM electron images of the strained samples heat treated at (**a**) 600 °C for 25 min; (**b**) 600 °C for 25 min in higher magnification; (**c**) 610 °C for 25 min; (**d**) 610 °C for 25 min in higher magnification.

It can be concluded from the above discussion that the distribution of alloying elements in the microstructure is significantly changed during the isothermal treatment and solution of the constituent particles. Segregation and distribution of the alloying elements has a great influence on the microstructural evolution during semi-solid treatment. Moreover, this affects the mechanical properties of the subsequent thixoformed parts [32,33,34]. Figure 6 shows the SEM images of the compressed samples heat treated at 600 °C (Figure 6a,b) and 610 °C (Figure 6c,d) for 25 min. The EDS analysis results of the marked areas (A to D) in Figure 6b are shown in Figure 7. Intense concentration of Cu at the grain boundaries (point A) was observed. In addition, Si intensely segregated at other regions of the grain boundaries (point B). The entrapped liquid droplets within the solid grains (point C) also showed high concentration of Cu. Although the segregation of Cu at the grain boundaries and within the liquid droplets has been reported in the literature [35,36], the segregation of Si is an unreported phenomenon in the semi-solid microstructure of 7075 Al alloy. Solid grains are depleted from Cu as a consequence of the segregation of Cu at the grain boundaries. This results in a rise in the solidus temperature in these regions and a decline of the temperature at the grain boundaries. Finally, chemical composition approaches Al-Cu eutectic composition which facilitates grain boundary melting. The segregation of Si at other regions of the grain boundaries derives grain boundary chemistry closer to Al-Si eutectic composition. Therefore, it can be concluded that the formation of low melting phases at grain boundaries is mainly influenced by the high content of Cu and Si.

Grain boundary regions with low solidus temperature remelt during the isothermal heating. Thus, a eutectic liquid film surrounds the solid grains. Melting of solid grains occurs on further increasing the heating temperature. However, it should be noted that the segregation of Cu and Si at the grain boundaries improves on raising the heating temperature due to more intense atomic diffusion. The EDS analysis of the precipitate particles at the grain boundaries (point D) in Figure 6 (Figure 7d) demonstrates Fe-rich intermetallic phases. This is also confirmed by XRD patterns in Figure 5. As discussed earlier, these particles cannot be dissolved at 600 °C and appear mainly as square particles at the grain boundaries in the semi-solid microstructure. The results of the microstructural investigations in this research showed that the amount of these particles reduce on increasing the isothermal heating temperature to 620 °C and isothermal heating for longer times (>15 min).

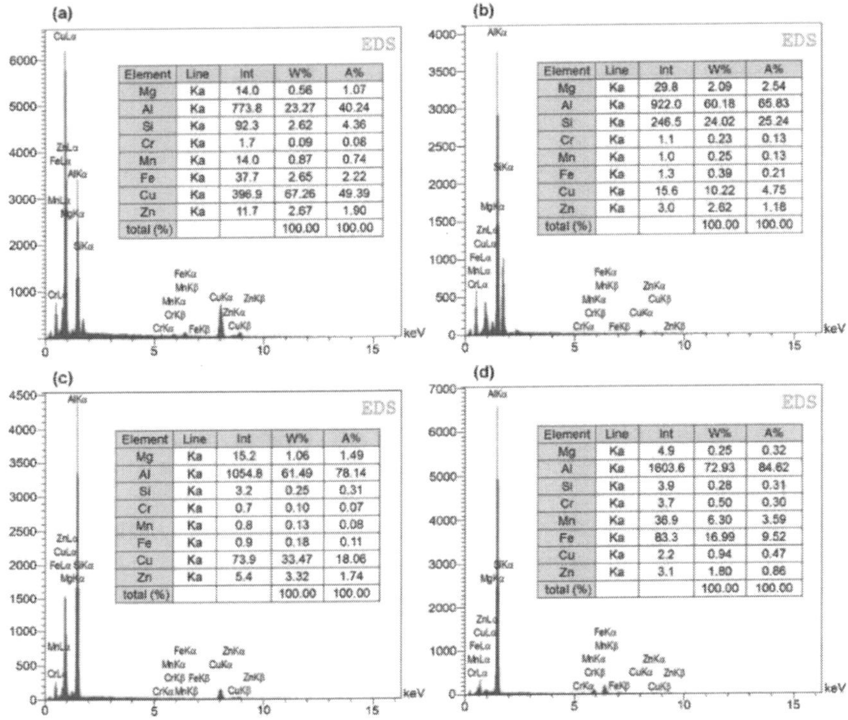

Figure 7. SEM/EDS analysis of various points in Figure 6: (**a**) A; (**b**) B; (**c**) C; (**d**) D.

Figure 8 shows the EDS mapping of the main alloying elements of the strained samples heat treated at 600 °C for 25 min. Considerable segregation of Cu and Si at the grain boundaries was also confirmed by the EDS mapping analysis of Cu and Si (Figure 8c,d). In addition, an appreciable depletion of Mg was observed at the grain boundaries (Figure 8e). However, as it can be observed from the EDS analysis results in Figure 7 and the EDS mapping of Mg in Figure 8e, the Al-Si eutectics have a higher content of Mg compared with the Al-Cu eutectics. There was no appreciable change in Zn content at various point of the microstructure (Figure 8f). This is in contrast to the result of researches conducted by Bolouri *et al.* [35] and Shim *et al.* [37] who reported the segregation of Zn at the grain boundaries of a SIMA processed 7075 alloy and semi-solid Al-Zn-Mg alloy prepared by the cooling plate method, respectively. Atkinson *et al.* [18] also indicated that Zn and Mg are enriched at the grain boundaries of RAP processed 7075 alloy. Moreover, as it can be observed in Figure 8g,h, the grain boundary precipitates contain a high amount of Fe and Mn. As the EDS analysis revealed (Figure 7c,d), these precipitates are attributed to Fe-rich intermetallic particles. An overview of the alloying elements distribution in the semi-solid microstructure of 7075 Al alloy can be observed in Figure 8i.

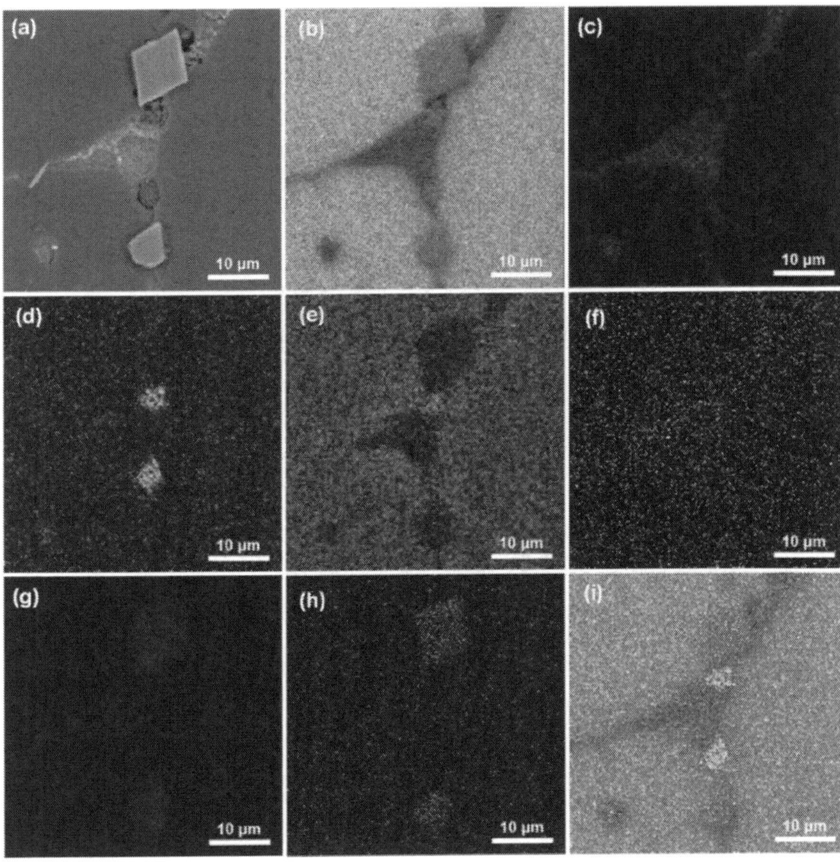

Figure 8. (**a**) SEM image; (**b**) Al; (**c**) Cu; (**d**) Si; (**e**) Mg; (**f**) Zn; (**g**) Fe; (**h**) Mn; (**i**) Al Fe Si Mn Cu X-ray image analysis of a compressed sample isothermally heat treated at 600 °C for 25 min.

Figure 9a–e shows SEM images of compressed samples heat treated at 600 °C for 5–35 min. Variations of the main alloying elements of 7075 alloy (*i.e.* Cu, Zn, and Mg) at the grain boundaries and grain centers in these samples are also demonstrated in Figure 9f. The EDS analysis results of the grain centers and grain boundaries (Figure 9f), showed that the Cu concentration reduces within the solid grains on increasing the isothermal holding time at 600 °C. In other words, Cu segregation rises at grain boundaries with increasing holding time. Mg concentration declined at the grain boundaries on raising the holding time and no significant change was observed in Zn concentration at various parts of the microstructure.

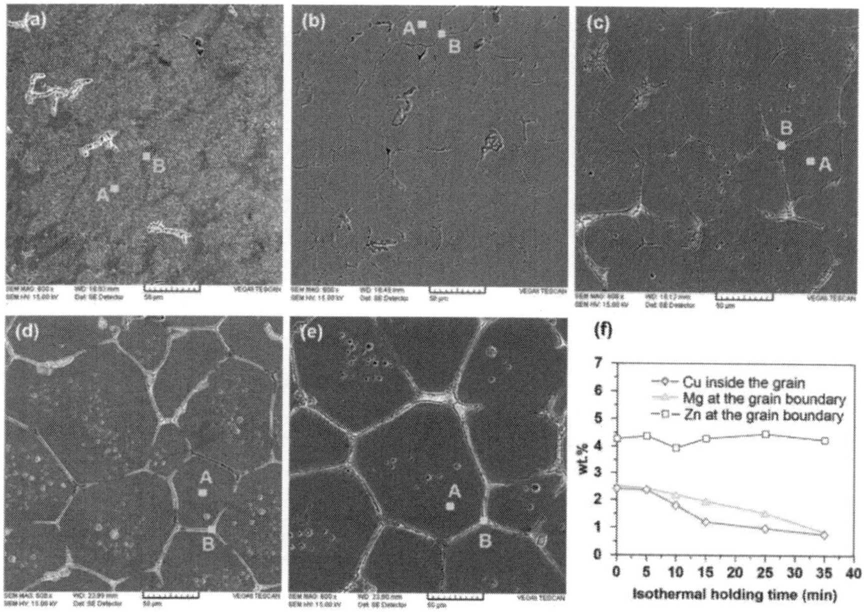

Figure 9. SEM images of sample compressed 20% , heat treated at 600 °C for (**a**) 5 min; (**b**) 10 min; (**c**) 15 min; (**d**) 25 min; (**e**) 35 min and then water quenched; (**f**) EDS analysis results of the grain centers (A point) and grain boundaries (B point).

Considering that (i) Zn, Mg, Si, and Cu-bearing intermetallic precipitates (η, S and Mg_2Si) are dissolved gradually on prolonging the holding time according to XRD patterns of the SIMA processed samples (Figure 5), (ii) Zn and Mg have higher solubility in Al compared with Cu and Si according to the Al-Zn, Al-Mg, Al-Cu and Al-Si binary phase diagrams [38], and (iii) no distinct peak for these elements appeared in the XRD patterns (Figure 5), it can be concluded that Zn and Mg elements are easily dissolved in α-Al, while Cu and Si significantly segregate at the grain boundaries. Since, Zn has a smaller atomic radius relative to aluminum (0.133 nm compare to 0.143 nm) [39], it can be concluded that dissolving Zn in the aluminum crystal lattice leads to reduction of the lattice parameter of aluminum. However, dissolving Mg in Al, as a result of its greater atomic radius, can result in increase of the lattice parameter of aluminum. The shifting of the Al main peaks in XRD patterns (Figure 5) to higher angles, as shown in Figure 10a for (111) reflection of the aluminum matrix, revealed that the lattice parameter of Al decreases on prolonging the heating time (Figure 10b) based on Bragg's equation ($n\lambda = 2d\sin\theta$). It seems that dissolving Zn in Al is more effective than Mg. This may be as a result of the higher Zn content of 7075 alloy. According to Figure 10b, the lattice parameter is gently reduced with prolonging holding time to 15 min and then with further extension of the holding time to 25 min, the lattice parameter precipitously declines. Finally, further increasing of the heating time results in a gentle reduction of the lattice parameter. These observations clearly correlate with the XRD results in Figure 5. According to the XRD patterns, great amounts of η precipitates dissolve during the isothermal heating of the sample for 25 min. This results in dissolving of a higher amount of Zn atoms in the aluminum lattice and therefore leads to a significant decline of the lattice parameter.

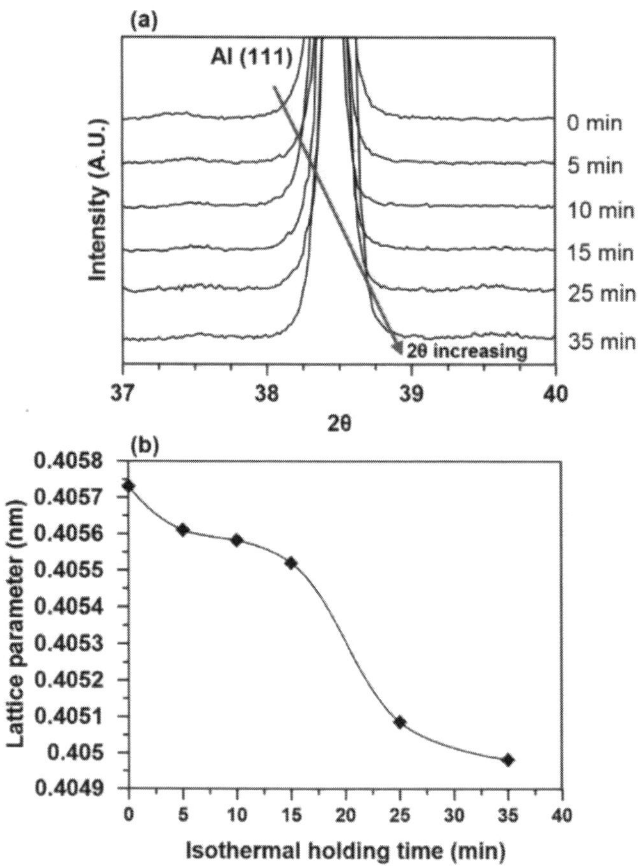

Figure 10. (**a**) XRD peaks corresponding to the Al(111) plane in XRD patterns of Figure 5; (**b**) lattice parameter variations of aluminum in 20% compressed samples heated at 600 °C for different times.

3.5. Effects of Isothermal Holding Temperature and Time on the Semi-Solid Microstructure

Figure 11 shows the quenched microstructures of 7075 samples which compressed 20% and heated at 600, 610, and 620 °C for different times. A close examination of the micrographs in Figure 11 reveals that the liquid fraction and the sphericity of solid grains become greater on raising the isothermal holding temperature and time. Another fact that can be observed in the micrographs is the reduction of the number of entrapped liquid droplets within the grains and increase of their size on extending the heating temperature and time.

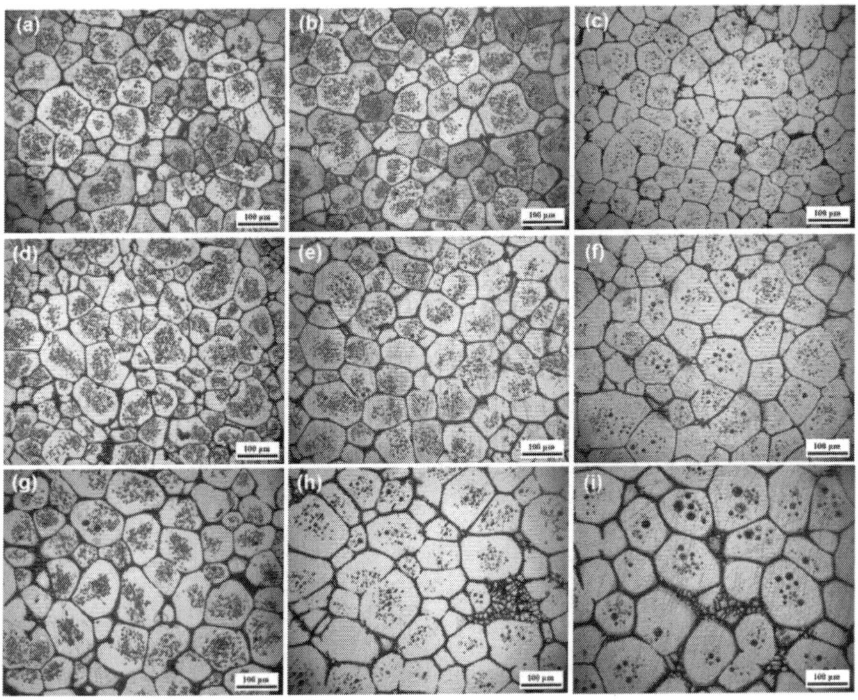

Figure 11. Optical micrographs of 7075 samples with 20% compression ratio heat treated at (**a**) 600 °C/15 min; (**b**) 610 °C/15 min; (**c**) 620 °C/15 min; (**d**) 600 °C/25 min; (**e**) 610 °C/25 min; (**f**) 620 °C/25 min; (**g**) 600 °C/35 min; (**h**) 610 °C/35 min; (**i**) 620 °C/35 min.

Variations of the average grain size and shape factor of strained samples heated at different temperatures and times are shown in Figure 12. As expected, the average grain size and shape factor become greater with increase of the holding temperature and time. However, it is worth noting that no significant variation of the average grain size of samples heated for 15 min on raising the holding temperatures from 600 to 620 °C is observed. Disintegration of solid particles owing to liquid phase penetration between the solid grains occurred during the primary heating time. This is due to the fact that a large number of solid grains with the same crystallographic orientations make contact with each other. It seems that the mechanism of particle disintegration is in equilibrium with the coalescence mechanism at longer heating times up to 15 min. Therefore, the solid grain sizes show no appreciable variation. In contrast, with further extension of holding times at various heating temperatures, the liquid film thickness between the solid grains and the liquid fraction becomes greater. Consequently, the Ostwald ripening mechanism can be considered as a dominant mechanism in the solid grain coarsening process. Figure 3e,f shows separated spherical grains with large entrapped liquid droplets which is a fact that supports the discussion above.

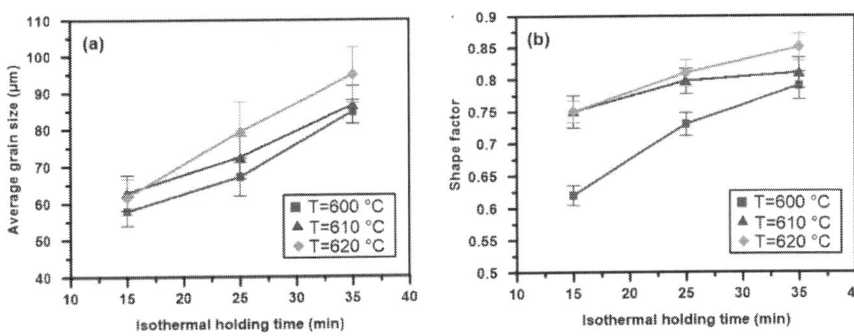

Figure 12. Variations of (**a**) the average grain size; (**b**) shape factor *versus* isothermal holding time of compressed samples heated at 600, 610, and 620 °C.

Results of research conducted by Vaneetveld *et al.* [40] showed that the semi-solid feedstock with high solid fraction ($f_s \approx 0.84$) shows better thixotropic behavior during semi-solid forming and also significantly prevents the formation of some defects such as porosity and shrinkage during the forming process. According to the solid fraction *versus* temperature curve in Figure 2b, the solid fraction is relatively low at 620 °C ($f_s \approx 0.55$). Thus, heating samples at such a temperature leads to undesirable coarsening of the solid grains. In addition, the solid fraction is more sensitive to temperature variations at 620 °C, when compared to the lower temperatures in which control of the semi-solid forming parameters is difficult. Consequently, the semi-solid microstructure obtained by isothermal heating at 620 °C is inappropriate for thixotropic applications. However, high solid fractions were observed following heating at temperatures of 600 and 610 °C. The solid volume fractions at 600 and 610 °C were 0.8 and 0.7, respectively (Figure 2b). Average grain size smaller than 75 μm and shape factor greater than 0.7 were obtained for strained samples heated at 600 and 610 °C for 25 min. Therefore, the isothermal heating temperature range of 600 to 610 °C and holding time of 25 min can be considered the optimum semi-solid treatment condition.

The coarsening kinetics can be expressed by the Lifshitz-Slyozov-Wagner (LSW) relationship [41,42]:

$$D^n - D_0^n = kt \tag{3}$$

where D and D_0 are the final and initial grain sizes, respectively, t is the isothermal holding time, k is the coarsening rate constant and n is the power exponent. It has been found that n is 3 for volume diffusion controlled systems in the semi-solid state [12].

In the present research, the coarsening rate constant (k) was calculated by fitting a power relationship to the experimental results. Figure 13 shows the cube of grain size variations *versus* isothermal holding time for compressed samples heated at 600, 610, and 620 °C, where D_0 is the average grain size when the holding time is 15 min. The regression coefficients of the fitted equations are close to 1. Thus, it can be concluded that the coarsening kinetics of solid particles during the isothermal heating of deformed 7075 samples at the semi-solid temperature range are fairly well correlated with the LSW equation. The values of the coarsening rate constant k for the samples are summarized in Table 3. The main characteristics of the variation of k values when the temperature was decreased from 620 to 600 °C are as follows; (i) the k value significantly reduced on decreasing the temperature from 620 to 610 °C; (ii) this was followed by a slight increase at 600 °C.

According to the Doherty theorem [32], the coarsening rate accelerates with rising solid fraction (f_s) for f_s higher than 0.6. However, the research conducted by Manson-Whitton *et al.* [41] showed opposite results for higher solid fractions ($f_s \geq 0.7$) in the case of spray formed Al-4% Cu. Therefore, they proposed a modified model considering the solid-solid contacts effect during coarsening and a transition solid fraction ($f_s \approx 0.7$). For the solid fractions greater than the transition value, the $k$value decreased with an increase in the solid fraction. However, for the solid fraction lower than the transition value, the $k$value increased on increasing the solid fraction. Considering the holding temperatures that correspond to the calculated solid fractions (Figure 2b), for solid fractions greater than 0.7 (at 610 °C) and the transition value of 0.8 (at 600 °C), the coarsening process of solid grains is similar to the model proposed by Manson-Whitton *et al.* [41]. On the other hand, an unexpected increase in k values is observed for solid fractions lower than 0.7. This can be attributed to (i) further increase of the atomic diffusion and more effective Ostwald ripening mechanism and (ii) a less retarding effect of the precipitate particles (Fe-rich intermetallic particles) for a holding temperature up to 620 °C. These particles bring about convoluting grain boundaries as is marked by arrows in Figure 14, suggesting pinning and retardation of the grain boundary liquid film migration during the solid grains coarsening. The effect of the presence of intermetallic precipitates at the grain boundaries on the coarsening rate of aluminum alloys has also been reported by Manson-Whitton *et al.* [41] and de Freitas *et al.* [43]. The results obtained in the present research indicate that Fe-rich precipitates mainly dissolve or become smaller than the thickness of the liquid film on raising the heating temperature up to 620 °C. Thus, the movement of grain boundaries can easily occur and leads to greater values of the coarsening rate constant.

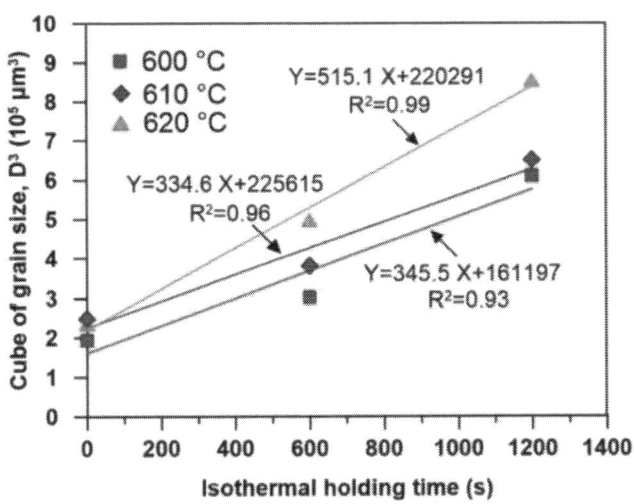

Figure 13. Variations of the cube of average globule size *versus* isothermal holding time of compressed samples at heating temperatures of 600–620 °C where R^2 is the regression coefficient.

Table 3. Values of coarsening rate constants of compressed samples at various isothermal holding temperatures.

Compression ratio (%)	Coarsening rate constant, k ($\mu m^3 \cdot s^{-1}$)		
	600 °C	610 °C	620 °C
20	345	334	515

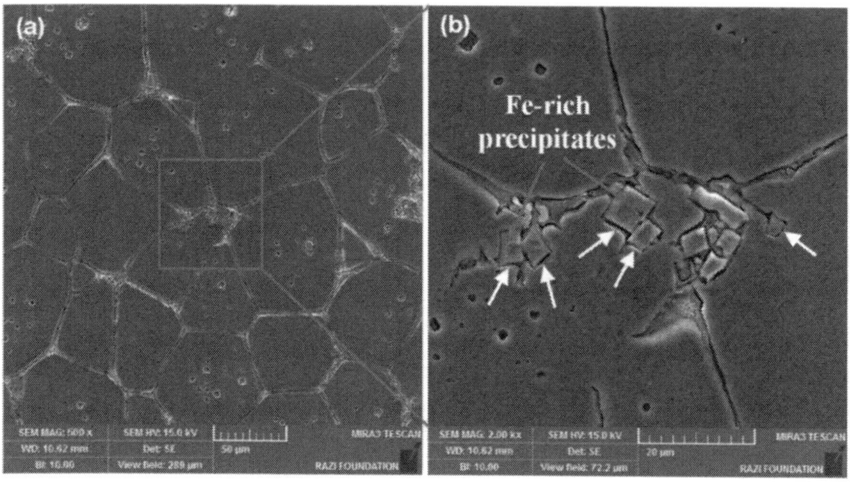

Figure 14. (**a**) SEM image of a strained sample heat treated at 600 °C for 15 min; (**b**) SEM image of selected area in (a).

3.6. Semi-Solid Deformation Behavior

Figure 15 shows a typical load-displacement curve obtained from the hot compression experiment. As it can be observed, the load rises to a maximum (L_{max}) and then decreases to a minimum value (L_{min}). The thixotropic flow behavior consisted of a transient regime and a steady state regime, as shown in Figure 15. Semi-solid alloys with an equiaxed microstructure are considered as deformable semi-cohesive granular solids saturated with liquid [44]. The solid grains are partially interconnected by unwetted grain boundaries and liquid fills the interstitial spaces. Large deformation results in the destruction of some solid grain boundary bonds to allow grains to move freely and rearrange through sliding and rolling. The cohesion at the grain level reflects the strength of unwetted grain boundaries and results in a flow resistance. With further deformation breakdown the solid agglomerates, the load (stress) decreases to a steady state level.

Tzimas and Zavaliangos [44] suggested that the flow resistance of semi-solid alloys at high solid volume fraction is controlled by four mechanisms: (i) elastoplastic deformation at grain contacts, (ii) destruction of cohesive bonds between solid grains, (iii) resistance to the flow of liquid relative to the solid, and (iv) resistance to grain rearrangement. Mechanism (i) is dominant in the rising part of the flow curve and mechanism (ii) results in the breakdown of solid particles and a significant drop in the load value. Mechanisms (iii) and (iv) are dominant in the second stage, and their relative contribution to the overall resistance to flow depends on the volume fraction of solid.

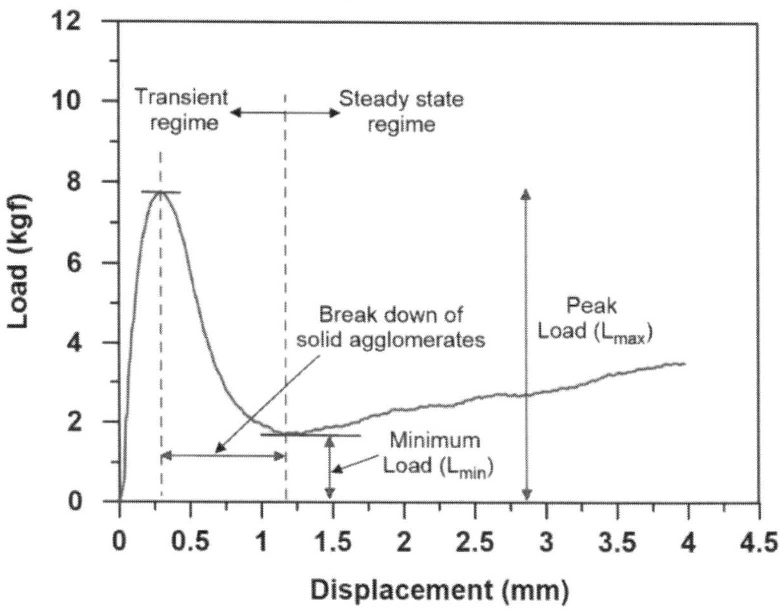

Figure 15. Typical load *versus* displacement curve of the hot compressed semi-solid 7075 sample (isothermally treated at 600 °C for 15 min) at 600 °C.

The stress-strain curves obtained from the SIMA processed 7075 samples with different initial microstructures after hot compression at 600 °C (for the SIMA samples isothermally treated at 600 °C for 10–35 min) and 610 °C (for the SIMA sample isothermally treated at 610 °C for 15 min) are shown in Figure 16. The stress exhibits a maximum value at a relatively low strain in all samples. The strain at the peak stress varies between 0.01 and 0.04. The peak stress then drops to a steady state value, typically at a strain of 0.1–0.18. The stress at the steady state shows no significant change with strain. Figure 17 shows the variation of the peak and steady state stresses for the SIMA processed samples with different initial microstructures. As is evident from Figure 16 and Figure 17, the sample semi-solid treated at 600 °C for 10 min exhibits the highest peak and steady state stresses. On increasing the isothermal holding time to 25 min at 600 °C, which results in a coarser and more spherical solid grains in the initial microstructure (see Figure 12), both the peak and steady state stresses decreased significantly. The peak stress showed a slight increase while a considerable increase in the steady state stress was observed with further increase of the holding time to 35 min in the SIMA process. Moreover, the increase of the deformation temperature had a great influence on the stress-strain curve and resulted in a decrease of thixotropic strength, as is observed for the sample semi-solid treated at 610 °C for 15 min.

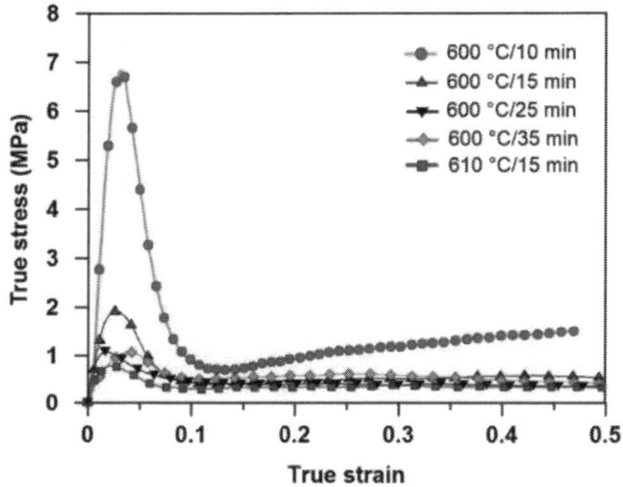

Figure 16. Stress-strain curves of strain induced melt activation (SIMA) processed samples with different initial microstructures.

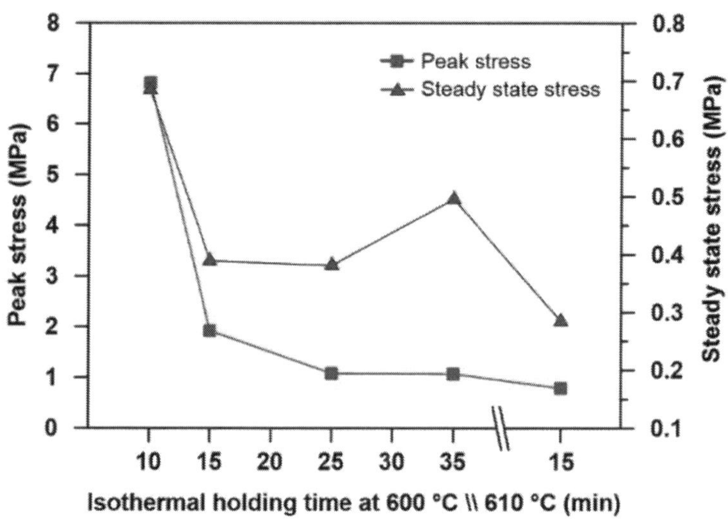

Figure 17. Peak and steady state stresses of the SIMA processed samples with different initial microstructures.

The descending trend of the thixotropic strength (decreasing of the peak stress) with increase of the isothermal holding time in the SIMA process can be attributed to the geometry of the solid grains. As was previously described, the deformation of samples with equiaxed microstructure mainly occurs by grain rearrangement through sliding and rolling. However, the microstructures consisting of non-equiaxed and elongated grains are deformed by plastic deformation due to which the relative movement among solid grains is constrained. Figure 18 shows the hot compressed microstructure of SIMA processed samples with near equiaxed and globular initial microstructure. The microstructure of various regions (1 to 3 in Figure 18a) of the hot compressed SIMA sample processed at 600 °C for 10 min (see Figure 3c) revealed that this sample cannot be deformed by grain rearrangement due to the significant geometric interference of the solid grains which results in plastic deformation of the grains (Figure 18b–d). As is observed in Figure 18b–d, the solid grains are severely deformed and elongated in the direction of compression which results in a textured microstructure. This leads to a much higher resistance to flow compared to samples with a completely spheroidized microstructure. On prolonging the isothermal heating time, the sphericity of solid grains increases and the interconnection decreases (Figure 3) which results in easier movement and sliding during deformation and reduction of flow resistance. The high flow stress of the SIMA sample heated for 35 min can be ascribed to the coarsened initial microstructure with an average grain size of 85 μm. According to Clarke [45], an increase in the particle size leads to an increase in the apparent viscosity of the semi-solid slurry. Therefore, a sample with coarsened initial microstructure possesses a more viscous flow and requires higher loads for thixoforming.

Figure 18e,f shows the microstructure of the center region (region 2 in Figure 18a) of the hot compressed semi-solid samples prepared by heating at 600 °C for 25 and 35 min, respectively. After compression, no evidence of the plastic deformation of the solid grains was observed in the microstructure of the samples. There was no difference in the microstructure of the center regions and edge regions, which indicated that the solid and the liquid flowed together during the compression. However, it is worth mentioning that the liquid fraction in the edge regions was slightly high, which was due to the liquid phase being squeezed out during the deformation. The microstructural observations also revealed that the solid grains do not experience plastic deformation in the SIMA samples prepared by heating at 600 and 610 °C for 15 min. According to Chen and Tsao [46], the plastic deformation of solid particles (PDS), sliding between the solid particles (SS), liquid flow (LF), and the flow of liquid incorporating solid particles (FLS) are four dominant mechanisms controlling deformation of alloys in the semi-solid state. The two former mechanisms are active when the solid particles are in contact with each other and the other two mechanisms are dominant when the solid particles are surrounded by the liquid phase. Therefore, it can be deduced that the sample with near equiaxed initial microstructure deform through the mechanism of PDS. In the case of globular initial microstructures, the SS and FLS mechanisms are dominant, however, the effectiveness of the SS mechanism may dwindle and the FLS mechanism is reinforced on prolonging the isothermal holding temperature and time due to the decrease of the interconnection between the solid grains.

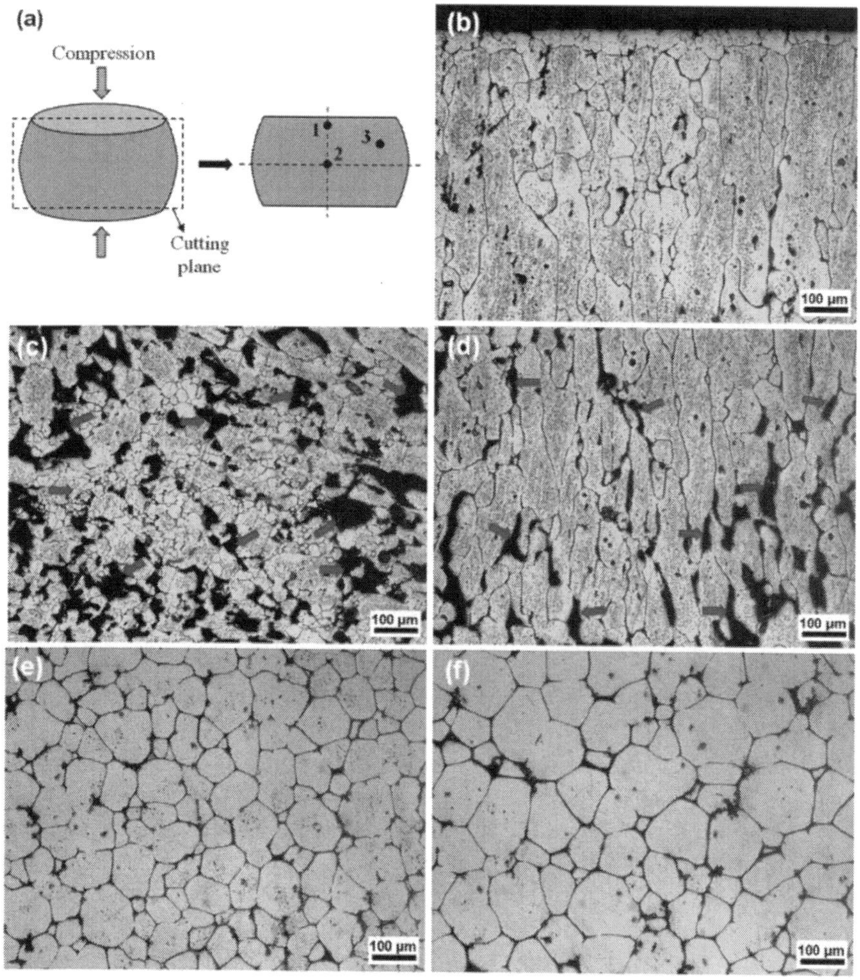

Figure 18. (**a**) Schematic of a hot compressed SIMA sample showing compression direction; Optical micrographs of (**b**) region 1; (**c**) region 2; (**d**) region 3 of SIMA sample processed at 600 °C for 10 min; region 2 of SIMA sample processed at 600 °C for (**e**) 25 min; (**f**) 35 min after hot compression at 600 °C.

The close examination of the hot compressed microstructure of different samples (Figure 18) also reveals that the microstructure of the SIMA processed sample with near equiaxed and elongated initial grains contains a higher amount of porosity, especially in regions 2 and 3, (shown by arrows in Figure 18c,d) compared to the samples with globular microstructure. In the sample with near equiaxed initial microstructure, the low amount of liquid content leads to higher viscosity, which may prevent the homogenous flow of the liquid and solid particles during deformation. Therefore, this sample shows a high amount of porosity in the microstructure. In addition, as shown by dotted ovals in Figure 18c, recrystallization occurs in some parts of the center region in the sample with an incomplete spheroidized initial microstructure during hot compression. It is believed that this recrystallization may occur dynamically as a result of higher induced strain in the center regions compared to the edge regions during hot compression.

4. CONCLUSIONS

The SIMA process consisting of applying uniaxial compression strain at ambient temperature and subsequent semi-solid treatment in the range of 600–620 °C for 5–35 min was investigated. It has been found that using the cold compression process following by heating in the semi-solid range results in some phase evolution that is suitable for obtaining a semi-solid microstructure for subsequent thixoforming. The following results can be drawn from the analysis:

The results showed that the partial remelting kinetics during the SIMA process consist of four steps including recrystallization and structural separation, coarsening of polygonal solid particles, spheroidization of polygonal particles, and coarsening of spherical particles. The growth and coarsening of the solid particles in the SIMA process are controlled by two mechanisms of coalescence (for low liquid fractions) and Ostwald ripening (for high liquid fractions).

The XRD results of the compressed samples heated at 600 °C for different times showed that $MgZn_2$ (η), Al_2CuMg (S), and Mg_2Si precipitates are dissolved gradually during isothermal heating through the phase transformations of α-Al + $\eta \rightarrow$ L and then α-Al + S + $Mg_2Si \rightarrow$ L. However, Fe-rich precipitates aggregate as square particles at grain boundaries due to their higher melting points.

An intense segregation of Si and Cu was observed at the grain boundaries in the semi-solid microstructure which results in shifting of the grain boundary composition toward the Al-Si and Al-Cu eutectics. In contrast, a significant depletion of Mg was observed at the grain boundaries and Zn distribution showed no appreciable change during isothermal treatment.

Microstructural observations indicated that the isothermal heating temperature range of 600–610 °C for 25 min can be considered an optimum condition for the SIMA process. Coarsening kinetics of the solid particles is fairly well correlated to the LSW theory during isothermal heating. Despite the higher liquid fraction of samples heated at 610 °C, the coarsening rate at 610 °C was lower than 600 °C.

Samples with near equiaxed initial microstructure containing elongated and polygonal solid grains and samples with globular initial microstructure with average grain size of 61 μm show the greatest and the lowest flow resistance during the thixoforming process at 600 °C, respectively. The flow resistance decreases on raising the deformation temperature.

AUTHOR CONTRIBUTIONS

Behzad Binesh performed the experiments and wrote the paper under Mehrdad Aghaie-Khafri's guidance, and contributed to all activities.

CONFLICTS OF INTEREST

The authors declare no conflict of interest.

REFERENCES

1. Kirkwood, D.H.; Suery, M.; Kapranos, P.; Atkinson, H.V. Semi-Solid Processing of Alloys; Springer: London, UK, 2009; pp. 109–112.
2. Flemings, M.C. Behavior of metal alloys in the semisolid state. Metall. Trans. A 1991, 22, 957–981.
3. Jayaraj, J.; Fleury, E.; Kim, K.-B.; Lee, J.-C. Globulization mechanism of the primary Al of Al-15Cu alloy during slurry preparation for rheoforming. Met. Mater. Int. 2005, 11, 257–262.
4. Zhang, Q.Q.; Cao, Z.Y.; Zhang, Y.F.; Su, G.H.; Liu, Y.B. Effect of compression ratio on the microstructure evolution of semisolid AZ91D alloy. J. Mater. Process. Technol. 2007, 184, 195–200.

5. Saklakoglu, N.; Saklakoglu, I.E.; Tanoglu, M.; Oztas, O.; Cubukcuoglu, O. Mechanical properties and microstructural evaluation of AA5013 aluminum alloy treated in the semi-solid state by SIMA process. J. Mater. Process. Technol. 2004,148, 103–107.

6. Chen, T.J.; Hao, Y.; Sun, J. Microstructural evolution of previously deformed ZA27 alloy during partial remelting.Mater. Sci. Eng. A 2002, 337, 73–81.

7. Tzimas, E.; Zavaliangos, A. A comparative characterization of near-equiaxed microstructures as produced by spray casting, magnetohydrodynamic casting and the stress induced melt activated process. Mater. Sci. Eng. A 2000, 289, 217–227.

8. Lin, H.Q.; Wang, J.G.; Wang, H.Y.; Jiang, Q.C. Effect of predeformation on the globular grains in AZ91D alloy during strain induced melt activation (SIMA) process. J. Alloys Compd. 2007, 431, 141–147.

9. Kang, B.K.; Hong, C.P.; Choi, B.H.; Jang, Y.S.; Sohn, I. Microstructural evolution in semisolid forging of A356 alloy.Met. Mater. Int. 2015, 21, 153–158.

10. Young, K.P.; Kyonka, C.P.; Kamado, S. Fine Grained Metal Composition. U.S. Patent 4415374, 30 March 1983.

11. Jiang, J.; Wang, Y.; Atkinson, H.V. Microstructural coarsening of 7005 aluminum alloy semisolid billets with high solid fraction. Mater. Charact. 2014, 90, 52–61.

12. Yan, G.; Zhao, S.H.; Ma, S.H.; Shou, H. Microstructural evolution of A356.2 alloy prepared by the SIMA process.Mater. Charact. 2012, 69, 45–51.

13. Jung, H.K.; Seo, P.K.; Kang, C.G. Microstructural characteristics and mechanical properties of hypo-eutectic and hyper-eutectic Al-Si alloys in the semi-solid forming process. J. Mater. Process. Technol. 2001, 113, 568–573.

14. Curle, U.A.; Govender, G. Semi-solid rheocasting of grain refined aluminum alloy 7075. Trans. Nonferrous Met. Soc. China 2010, 20, s832–s836.

15. Yong, L.S.; Hwan, L.J.; Seon, L.Y. Characterization of Al 7075 alloys after cold working and heating in the semi-solid temperature range. J. Mater. Process. Technol. 2001, 111, 42–47.

16. Chayong, S.; Atkinson, H.V.; Kapranos, P. Thixoforming 7075 aluminum alloys. Mater. Sci. Eng. A 2005, 390, 3–12.

17. Chayong, S.; Atkinson, H.V.; Kapranos, P. Multistep induction heating regimes for thixoforming 7075 aluminum alloy. Mater. Sci. Technol. 2004, 20, 490–496.

18. Atkinson, H.V.; Burke, K.; Vaneetveld, G. Recrystallisation in the semi-solid state in 7075 aluminum alloy. Mater. Sci. Eng. A 2008, 490, 266–276.

19. Vaneetveld, G.; Rassili, A.; Atkinson, H.V. Influence of parameters during induction heating cycle of 7075 aluminum alloys with RAP process. Solid State Phenom. 2008, 141–143, 719–724.

20. Bolouri, A.; Shahmiri, M.; Kang, C.G. Study on the effects of the compression ratio and mushy zone heating on the thixotropic microstructure of AA 7075 aluminum alloy via SIMA process. J. Alloys Compd. 2011, 509, 402–408.

21. Mohammadi, H.; Ketabchi, M.; Kalaki, A. Microstructure evolution of semi-solid 7075 aluminum alloy during reheating process. J. Mater. Eng. Perform. 2011, 20, 1256–1263.

22. Seo, P.K.; Kang, C.G. The Effect of raw material fabrication process on microstructural characteristics in reheating process for semi-solid forming. J. Mater. Process. Technol. 2005, 162–163, 402–409.

23. Zhang, L.; Liu, Y.B.; Cao, Z.Y.; Zhang, Y.F.; Zhang, Q.Q. Effects of isothermal process parameters on the microstructure of semisolid AZ91D alloy produced by SIMA. J. Mater. Process. Technol. 2009, 209, 792–797.

24. Doherty, R.D.; Lee, H.-I.; Feest, E.A. Microstructure of stir-cast metals. Mater. Sci. Eng. 1984, 65, 181–189.

25. Tzimas, E.; Zavaliangos, A. Evolution of near-equiaxed microstructure in the semisolid state. Mater. Sci. Eng. A 2000,289, 228–240.

26. Wang, T.; Yin, Z.-M.; Sun, Q. Effect of homogenization treatment on microstructure and hot workability of high strength 7B04 aluminum alloy. Trans. Nonferrous Met. Soc. China 2007, 17, 335–339.

27. Fan, X.-G.; Jiang, D.-M.; Meng, Q.-C.; Zhang, B.-Y.; Wang, T. Evolution of eutectic structures in Al-Zn-Mg-Cu alloys during heat treatment. Trans. Nonferrous Met. Soc. China 2006, 16, 577–581.

28. Lim, S.T.; Lee, Y.Y.; Eun, I.S. Microstructural evolution during ingot preheat in 7xxx aluminum alloys for thick semiproduct applications. Mater. Sci. Forum 2006, 519–521, 549–554.

29. Li, N.-K.; Cui, J.-Z. Microstructural evolution of high strength 7B04 ingot during homogenization treatment. Trans. Nonferrous Met. Soc. China 2008, 18, 769–773.

30. Jiang, L.-T.; Wu, G.-H.; Yang, W.-S.; Zhao, Y.-G.; Liu, S.-S. Effect of heat treatment on microstructure and dimensional stability of ZL114A aluminum alloy. Trans. Nonferrous Met. Soc. China 2010, 20, 2124–2128.

31. Boettinger, W.J.; Kattner, U.R.; Moon, K.W.; Perepezko, J.H. DTA and heat-flux DSC measurements of alloy melting and freezing. In Methods for Phase Diagram Determination, 1st ed.; Zhao, J.C., Ed.; Elsevier: Ames, IA, USA, 2007; pp. 194–200.

32. Annavarapu, S.; Doherty, R.D. Inhibited coarsening of solid-liquid microstructures in spray casting at high volume fractions of solid. Acta Metall. Mater. 1995, 43, 3207–3230.

33. Liu, D.; Atkinson, H.V.; Kapranos, P.; Jones, H. Effect of heat treatment on properties of thixoformed high performance 2014 and 201 aluminum alloys. J. Mater. Sci. 2004, 39, 99–105.

34. Cavaliere, P.; Cerri, E.; Leo, P. Effect of heat treatments on mechanical properties and fracture behavior of a thixocast A356 aluminum alloy. J. Mater. Sci. 2004, 39, 1653–1658.

35. Bolouri, A.; Shahmiri, M.; Kang, C.G. Coarsening of equiaxed microstructure in the semisolid state of aluminum 7075 alloy through SIMA processing. J. Mater. Sci. 2012, 47, 3544–3553.

36. Jiang, H.; Li, M. Microscopic observation of cold-deformed Al-4Cu-Mg alloy samples after semi-solid heat treatments.Mater. Charact. 2005, 54, 451–457.

37. Shim, S.-Y.; Kim, D.-H.; Chio, S.-H.; Kim, Y.-H.; Bai, H.; Lim, S.-G. The analysis of variance on process parameters affecting the microstructures of semi-solid Al-Zn-Mg alloy billet by cooling plate method. Met. Mater. Int. 2010, 16, 1009–1017.

38. Baker, H. ASM Handbook, Alloy Phase Diagrams; ASM International: Materials Park, OH, USA, 2004; Volume 3, pp. 295–331.

39. Shackelford, J.F.; Alexander, W. Materials Science and Engineering Handbook, 3rd ed.; CRC Press LLC: Boca Raton, FL, USA, 2001.

40. Vaneetveld, G.; Rassili, A.; Pierret, J.C.; Beckers, J.L. Conception of tooling adapted to thixoforging of high solid fraction hot-crack-sensitive aluminum alloys. Trans. Nonferrous Met. Soc. China 2010, 20, 1712–1718.

41. Manson-Whitton, E.D.; Stone, I.C.; Jones, J.R.; Grant, P.S.; Cantor, B. Isothermal grain coarsening of spray formed alloys in the semi-solid state. Acta Mater. 2002, 50, 2517–2535.

42. Li, P.; Chen, T.; Zhang, S.; Guan, R. Research on semisolid microstructural evolution of 2024 aluminum alloy prepared by powder thixoforming. Metals 2015, 5, 547–564.

43. De Freitas, E.R.; Ferracini, E., Jr.; Ferrante, M. Microstructure and rheology of an AA2024 aluminum alloy in the semi-solid state, and mechanical properties of a back-extruded part. J. Mater. Process. Technol. 2004, 146, 241–249.

44. Tzimas, E.; Zavaliangos, A. Mechanical behavior of alloys with equiaxed microstructure in the semisolid state at high solid content. Acta Mater. 1999, 47, 517–528.

45. Clarke, B. Rheology of coarse settling suspensions. Trans. Inst. Chem. Eng. Chem. Eng. 1967, 45, 251–256.

46. Chen, C.P.; Tsao, C.-Y.A. Semi-solid deformation of non-dendritic structures-phenomenological behavior. Acta Mater.1997, 45, 1955–1968.

CHAPTER 8

Aluminium Alloys in Solar Power – Benefits and Limitations

Amir Farzaneh[3], Maysam Mohammadi[4], Zaki Ahmad[1, 2] and Intesar Ahmad[5]

[1] KFUPM. Dhahran, Saudi Arabia

[2] Dept Of Chemical Engineering, Comsats Lahore, Pakistan

[3] Dept of Materials Engineering, University of Tabriz, Tabriz, Iran

[4] Department of Materials Engineering, University of British Columbia, Vancuver, B.C, Canada

[5] Department of Electrical Engineering, Lahore College For Women University, Lahore, Pakistan

1. INTRODUCTION

It has been widely accepted that fossil fuels have played significance roles in human's daily life and industrial developments. However, fossil fuels are associated with some problems like air pollution. Moreover, fossil fuels have limited sources in the world that will be finished in near future if the consuming rate of fossil fuels does not decrease. So, scientists have been encouraged to find suitable sources of energy as replacements for fossil fuels.

Sunlight, wind, see waves and hydrogen are examples of renewable sources of energy that have been subjects of researches for many years.

Among all of he mentioned renewable sources of energy, the sun is the most abundant and most renewable source that provides the energy for human much more than they need. If we can collect 0.01% of the sun's energy reaching the earth, it would be more than energy that all human use today.

Amount of the energy that reaches from the sun to a specific location on the surface of the earth can be 1000 W/m^2 [1]. Solar energy is had been received great world wide attention during the last decades as the most ideal renewable source of energy, which is mainly due to the points that this energy is safe, clean, free and unlimited [2]. Basic of fusion reaction in the sun and solar energy production have been discussed in many monographs [3-6].

The sun sends out huge quantity of electromagnetic radiations to the earth, which transfer approximately 4000 trillion kWh of energy to the earth surface each day [7]. This amount of energy is much higher than human's demand and also more than any other source of energy on the earth like nuclear power or fossil fuel burning. Solar energy is associated with some economic and environmental advantages, and has been considered as replacement source of energy for traditional sources of energy [8].

History of employing solar energy goes back to 212 BC, when Archimedes. The Greek scientist used metallic mirror to burn a Roman fleet [9]. Cooking food, heating water and home, and drying grains were the first applications of the sun's energy [10-12]. In 18th century, solar furnaces were constructed by glass lenses, mirrors and polished iron, which were capable to melt metals like iron and copper. During 19th century, solar energy was used to operate steam engine and convert solar to electrical energy [13].

Nowadays, solar energy are used in wide range of industrial, business and residential applications such as electricity generation, water heating, industrial processes, daylighting, heating and cooling [1]. As a consequence of rapid development of the solar power technologies, it is expected that solar systems will provide 12% to 25% of global electricity by 2050 [8].

Growing demand for heat and electricity generation in developed and developing countries causes rapid development in solar power generation systems. Quality and physical properties of materials that are used in solar systems determine the efficiency of each solar system. Different materials are used in various kinds of solar power systems such as glass, silver, steel, stainless steel and aluminium.

Among all of the mentioned materials, aluminium has special properties that make it an interesting material for many solar power companies. Light weight, high strength, proper corrosion properties, high surface reflectivity, excellent electrical and thermal conductivities, as well as special optic properties of its anodic coating are such as interesting properties of aluminium that make it inseparable part of solar power systems.

To sum up, aluminium plays an important role in various kinds of solar power systems include concentrating solar power (CSP), photovoltaic solar power (PV) and solar thermal collections. The application of aluminium and its alloys in these solar systems are explained in this chapter. Besides, its economical effect and future market are explored here.

2. ALUMINIUM APPLICATIONS IN SOLAR POWER SYSTEMS

In order to find the role of aluminium and its alloys in solar power systems, it is necessary to review different types of solar power plants, their properties, requirements and applications. Generally, solar power systems are divided into three widely used categories, which called concentrating solar power (CSP), solar thermal absorbers and photovoltaic solar cells (PV). Aluminium alloys have became a significant and inseparable part of each of the mentioned group of solar power systems, mainly due to special properties of aluminium and its alloys. Properties and applications of each kind of the mentioned solar power systems as well as the role of aluminium alloys in each of them will be discussed separately.

2.1. Concentrating Solar Power (Csp)

Concentrating solar power systems include reflector materials that concentrate heat energy .of the sun to a point or line to generate steam in a boiler, drive steam turbine and produce electricity [14-36]. Generation electricity, however, is not the only application of the CSP systems. Concentrated solar energy is suitable source of energy that can be used in wide range of materials processing such as producing metallic foams [37], synthesis of nanomaterials like carbon nanotubes and ceramic nanoparticles [38], fast heating of ceramic materials close to thermal shock [39], breaking down metal oxides to its metal counterpart [40-42], surface treatment [38] and metal sintering [43].

Cost of energy generation by CSP is much lower than that of PV, which is mainly due to higher average efficiency of CSP (42% compared with 15% for PV) and requires smaller field to produce certain amount of energy [8,44]. The cost of energy generation by solar collection devices is also lower than thermal solar collecting systems[45].

CSP has facilitated a system for production of energy which is neither noisy nor toxic. Therefore, this system can be used in cities without any safety problem [2,44].

Based on the focus geometry and receiving technology, CSP has been divided into four types.

2.1.1. Parabolic Trough

In this kind of CSP, a series of curved mirrors concentrate sunlight on to a tube that located in the trough's focal line. This tube contains oil and is used as heat transfer medium. Temperature of the oil in the tube can reach as high as 400°C [46]. Figure1 shows a real and a schematic of a parabolic trough CSP system. This system was used in 1912 in Egypt for the first time [8]. The parabolic trough system is able to concentrate sunlight by 70-100 times and transfer solar to electrical energy with efficiency of 15% [47].

Figure 1. Parabolic trough concentrating solar power plant [47]

2.1.2. Solar Tower

Solar tower plants use many reflectors called heliostats to collect sunlight reaching a field to a certain point on top of a tower, where a collector is located (Figure 2). The concentrated energy can generate electric energy with efficiency of 20-35% [8,46].

2.1.3. Parabolic Dish

Parabolic dish concentrating solar system uses a reflector dish and concentrate sunlight onto its focal point. Figure 3 shows the schematic of this solar system. A receiver that located in the focal point of the dish can increase the temperature of the gas or fluid up to 750°C in a sunny day. Capacities of parabolic dish plants and their efficiency in changing solar to electrical energy are in the range of 0.01-0.4 MW and 25-30%, respectively.

Due to its design, its optical efficiency is significantly higher than two other mentioned categories of CSP plants. Parabolic dishes are currently used in some simple application like cooking oven [8,13,46].

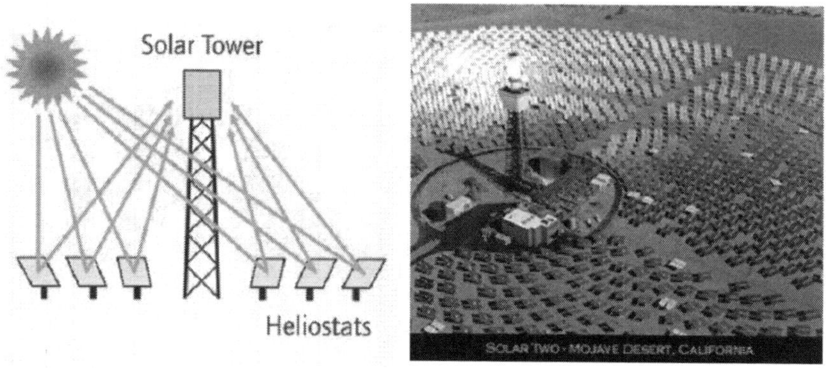

Figure 2. Solar tower concentrating solar power plant [47].

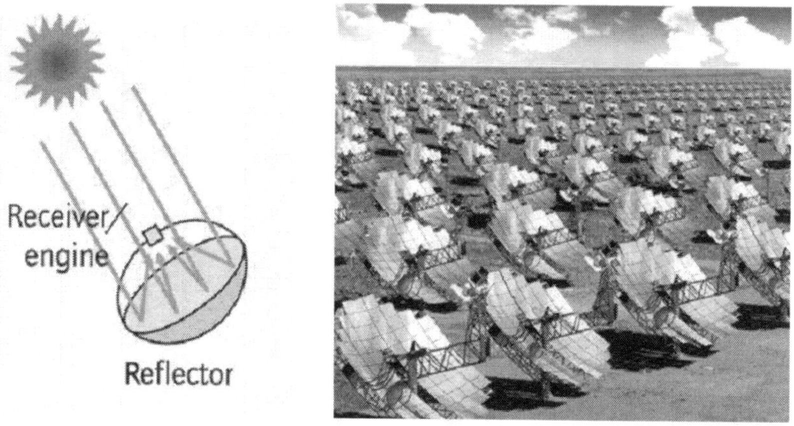

Figure 3. Parabolic dish concentrating solar power plant [47].

2.1.4. Linear Fresnel

Linear Fresnel is a line-focus collector that consists of series of Fresnel lens behaving mirrors (Figure 4). Capacity of this solar concentrator is 10-200Mw and its efficiency to generate electric energy is between 8 and 10% [8].

Table 1 compares various properties of the mentioned CSP plants.

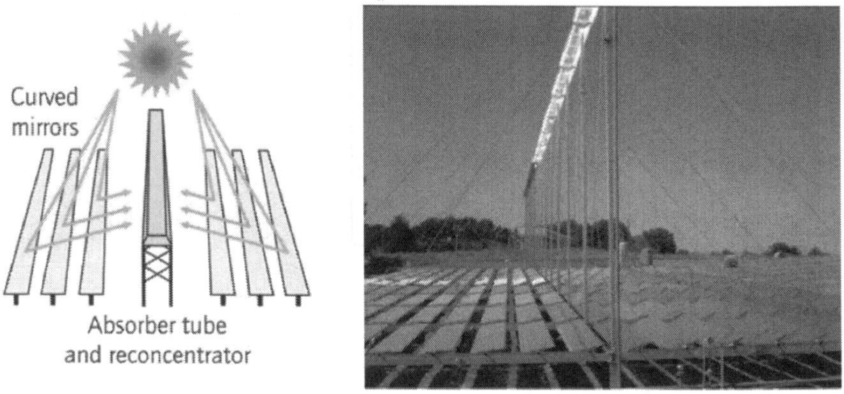

Figure 4. Linear Fresnel concentrating solar power plant [47].

Table 1. Properties of CSP technologies [8,46,47]

	Capacity (MW)	Concentration	Annual solar to electric conversion efficiency	Cost (Dollars /ft² field)
Parabolic trough	10-200	70-80	15%	40
Solar tower	10-200	300-1000	20-35%	45
Parabolic dish	0.01-0.4	1000-3000	25-30%	
Linear Fresnel	10-200	25-100	8-10%	20

Currently, capacity of installed concentrated solar power plants is 516MW in the world. 90.2% of this capacity belongs to parabolic trough systems, followed by 8.5% for solar tower, 0.8% for Fresnel trough systems and 0.24% for solar dishes [48].

Reflectors of concentrated solar power should have some properties, such as high reflectance ratio and low absorption, good wetability and low cost. Aluminium and silver are the most common reflectors for CSP systems due to their high reflecting efficiency in solar wavelength range [45].

Reflectors of concentrated solar power should have some properties, such as high reflectance ratio and low absorption, good wetability and low cost. Aluminium and silver are the most common reflectors for CSP systems because of their high reflecting efficiency in solar wavelength range [45].

Aluminium has some special properties that make it a useful mirror in various applications of solar cells, lasers and astronomer's instruments [2].

For example, aluminium can be deformed easily to have the best shape of reflectors and achieve the highest concentrating efficiency. Unlike glass mirrors, aluminium reflectors can not broken easily, which is a favourite property for outdoor applications [49].

Aluminium mirrors not only have better surface reflectivity than glass mirrors, they are much lighter. Compared to glass mirrors that have average weight of 11kg/m^2, aluminium reflectors have only weight of 7 kg/m^2.

Due to mechanical properties of aluminium and its low cost compared with silvered glass mirrors, aluminized reflectors found applicability to high temperature solar concentrating technologies [50].

Rolled aluminium also can be suitable for certain solar energy applications since it is cheaper than other reflector materials and can be cost-effective material in this application [45].

Thermal evaporation is one of the most practical methods to prepare aluminium reflector in order to use in concentrated solar power systems [2].

Ling et al. [2] studied performances of aluminium reflectors produced by thermal evaporation method on different substrates include galvanized iron, acrylonitrile butadiene styrene (ABS) and aluminium alloy. Experimental results clarified that reflection of thermally evaporated aluminium on ABS is comparable with that of silver mirror of ultra-white glass. It was also found that smoothness and roughness of the substrate have important effects on optical properties of the aluminium reflectors.

2.2. Solar Thermal Collectors

Solar thermal collector is a kind of solar power system that transforms solar energy from the sun rays into thermal energy. This solar system is widely used for generation of hot water, home heating and electricity generation [1,48].

Based on the kind of used collectors, solar thermal collecting systems divided into three types [1,13,48]:

2.2.1. Unglazed Plastic Collectors

This technology of solar thermal collectors provides a low cost heat and is used only for public bath heating and hay dying. Aluminium and its alloys have approximately no special application in unglazed plastic collectors.

2.2.2. Flat-Plate Collectors

Lower radiation and convection heat transfer losses in this solar thermal collectors compared to unglazed plastic collectors enables it to warm up to higher temperature. This system is widely used for space heating and hot water generation.

2.2.3. Evacuated Tube Collectors

Due to special design of this kind of solar thermal collectors, its conductive and convective losses are very low. So, it is capable to warm up the heat carry fluid up to 150°C.

According to figures recently released in Solar Heat World-wide report, the amount of energy that was generated by solar thermal collectors in the world in 2007 was approximately 147 GW. Fabrication of this amount of energy by solar thermal collectors requires approximately 210 square kilometres. Solar heating capacity of these solar systems had increased by 15% in the year of 2008, and became double compared to 2004. Capacity of solar thermal collectors does not equally divided among the mentioned solar thermal categories. According to published figures, approximately 50% of total amount of energy resulted from solar thermal collectors has been generated Evacuated tube collectors. Contributing proportions of Flat plate and Unglazed plastic collectors are 33% and 17%, respectively.

As mentioned before, Unglazed plastic collectors do not provide significant opportunity for aluminium usage. However, both of the other groups use aluminium and its alloys in different parts.

Figure 5 shows a Flat-plate collector and introduces its various parts; casing, absorber and frame. Aluminium, copper and steel are the materials that are used for absorbers. Casing and frames are usually made of aluminium and steel. However, aluminium is predominantly used because of lower weight of this alloy in that that of steel. Moreover, using aluminium as absorbers is growing. Special optic properties of anodic layer of aluminium and some aluminium alloys make aluminium a useful material for solar absorption. These qualities will be explained later. Today, approximately 35% of the solar absorbers are made of aluminium [1,13,48].

Figure 6 shows main components of Evacuated tube solar thermal collectors, which are absorber, frame, heat pipes, header pipe and casing. Like what was mentioned for flat-plate collectors, using aluminium as absorber is growing. Low density of aluminium satisfies solar companies to use aluminium alloys for frames instead of stainless steel. Aluminium is also widely used in casing and header pipes [1,13,48].

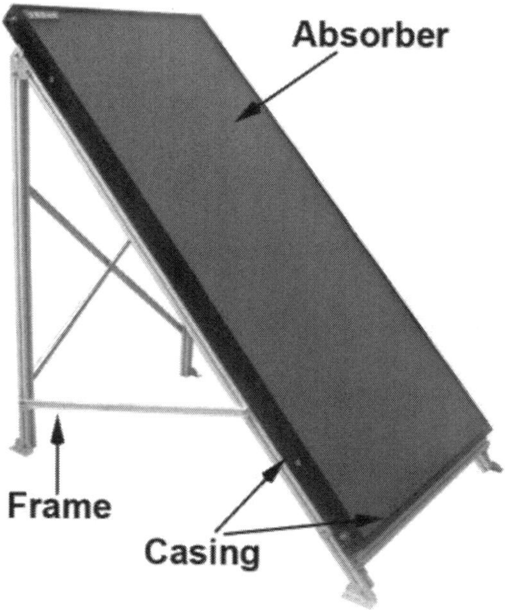

Figure 5. Different parts of a Flat-plate collector

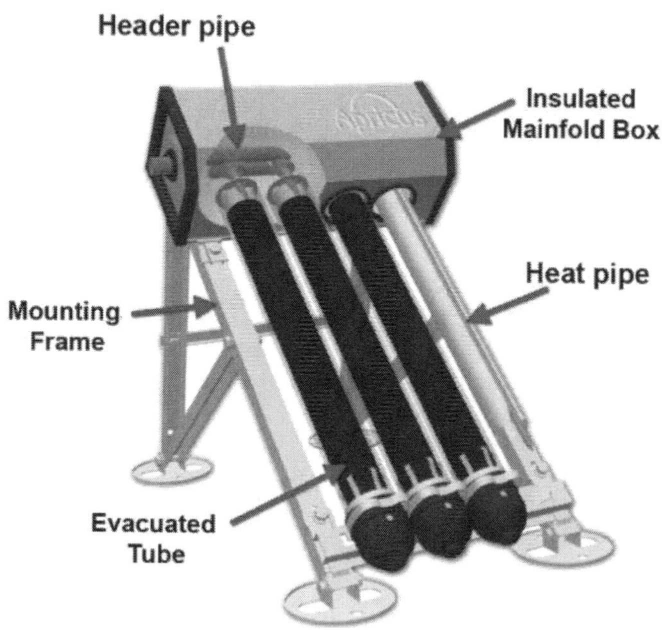

Figure 6. Different parts of a Evacuated tube solar thermal collectors

As mentioned before, aluminium is one of the most important materials utilized in solar absorbing, which is mainly due to special structure of the anodic layer that can fabricated on aluminium surface by anodizing process. This anodic layer is porous; with various pore sizes depend on the kind of anodizing electrolyte and anodizing process conditions [51-56].

It has been shown that these pores can be filled with metals by electrodeposition to have a coloured surface layer composed of metal particles in alumina dielectric matrix [57,57,58]. Using some special metal particles or ions, like nickel, provides suitable optical properties in anodized layer of aluminium for solar absorption application [57,59]. It was found that surface reflection of coloured aluminium anodized layer, as an undesirable property of solar absorbers, is a function of colouring time. It also was shown that reflectance decreases with electrolytic coloration time [57].

The effect of coloring time on solar absorption of aluminium anodized layer is more obvious in thicker anodic layers [53].

The proposed model of coloured and sealed anodized aluminium layer is shown in Figure 7. This model shows that the aluminium anodic layer composed of three parts, which have different properties. The top region (part 1) is protective layer that composes of Al_2O_3 and aluminium hydroxide; the middle region (part 2) is optical absorption layer that filled with metallic pigments; and the lowest region (part 3) is barrier layer that is a compact layer and has the highest density among all of the mentioned parts [57].

Absorption mechanism of solar radiation in colored anodic aluminium layer has been described by Granqvist [58].

For a photothermal conversion, the highest efficiency will be provided when the thermal losses by surface radiation are low enough, and solar absorption is high.

Microstructure of an alloy plays an important role on its efficiency as an absorber. As shown by Cody and Stephens [60], surfaces that its dielectric constant change gradually from the material/air interface to that of solid materials have low reflections.

Eutectic binary aluminium alloys such as Al-6wt.% Ni, Al-33wt.% Cu and Al-7.5wt% Ca have such microstructure and have acceptable optic properties to be used as absorber; i.e. low reflection high absorption [61,62].

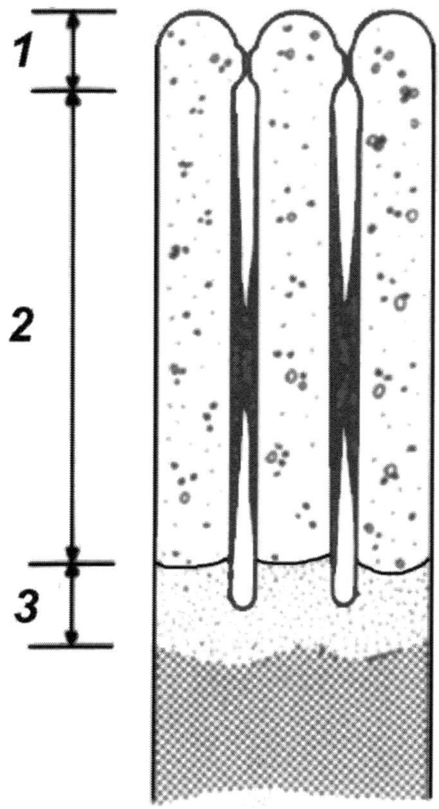

Figure 7. Proposed model of coloured and sealed anodized aluminium layer [57].

Mechanical and thermal stability of the mentioned aluminium alloys, as well as possibility of regeneration of their surface by retching make them useful materials in this application [61].

2.3. Photovltaic Solar System

Photovoltaic cells directly convert solar to electrical energy using semiconductor materials. Semiconductors can generate free electrons using energy of sunlight [63]. Photovoltaic property of materials had been discoveres by Becquerel in 1830, when he found this effect in Selenium [13]. Various aspects of photovoltaic solar systems have been reported in different books and references [64-72].

Apace was the first application of photovoltaic solar cells because sun is the only source of energy in space [13]. Photovoltaic solar cells have been used in wide range of applications include water pumping, solar home systems, remote building, solar cars and airplanes, satellites and space vehicles. Such a vast variety of applications is the main cause for increasing demand for photovoltaic solar cells [63].

Photovoltaic solar power system has experienced great development rate. The capacity of this solar system had increased from 0.1 GW in 1992 to 2.8 GW in 2007 and 5.95 GW in 2008. Table 2compares the market demand of PV in some countries in 2007 and 2008 [1].

Table 2. Photovoltaic solar system market in different regions of the world [1]

Countries	USA	Japan	Germany	Spain	Rest of Europe
Demand in 2007 (MW)	226	226	1328	650	170
Demand in 2007 (MW)	360	230	1860	2460	310

Various advantages of photovoltaic solar systems make it a favourite technology in many industries. It has been observed that PV is one of the fastest growing industries in the world. This rapid growth requires new developments with respect of applicable structural materials [73].

Construction and structure of photovoltaic solar systems are the main part of this system that can be made of aluminium. Steel and aluminium are the most common materials that are used in construction of solar power systems.

However, the advantages of aluminium alloys over steel, other aluminium alloys and composite materials make it the core material in building of large scale solar generation fields. Significant proportion of the cost of solar generation system is related to supporting materials and frames. For instance, approximately 25-30% of the budget of CSP plants should be allocated to frames and support materials.

Some of the aluminium alloys can perform same as steel, but its weight is one-third of steel. Although aluminium alloys are more expensive than steel, using aluminium instead of steel can be economical. Lower density of aluminium gives the opportunity of easier, faster and cheaper transportation. Moreover, construction of a solar field would be faster if extruded aluminium is used instead of steel because it does not need crane and permanent joining process like welding. These properties of aluminium enable engineers to design and produce complex, efficient and stable structures.

6061 aluminium alloy that contains magnesium and silicon alloying elements is an example of useful aluminium alloys for structure of solar plants. This aluminium alloy is widely used in solar fields because of its high strength and machinability[74].

Another advantage of aluminium over steel is its higher corrosion resistance in outdoor environments, even if steel is galvanized. Even though aluminium is more chemically and electrochemically active than steel, a thin oxide layer that naturally formed on the aluminium surface in the air provides suitable protection for aluminium and enables it to have good performance for a long time. To illustration, consider the aluminium cap used in Washington monument was corroded only 0.13 mm after 73 years performance. The mentioned oxide layer is stable in pH range between 4.5 and 8.5. So, it may not provide enough protection if the aluminium is located in soil. However, most of solar collectors are mounted on concrete pads, which make the aluminium performance independent of soil conditions [74].

Since using solar power for electricity generation has became a serious competition among various companies, designing and using the most possible available, effective and efficient materials became very important. Extruded aluminium can be considered as one of these effective materials as it enables companies to create next generations of solar power plants with long life time and very low negative environmental effects.

PV inverter, which changes direct current to alternative current, and panel frame are the other components of a photovoltaic solar system that can be made of aluminium

Approximately 72% of aluminium input in photovoltaic solar systems is used in construction, while the proportion of aluminium used in panel frames and inverters are 22% and 6%, respectively [48].

2.4. Perspective Of Aluminium Applications In Solar Power Systems

Currently, CSP systems use approximately 55000 kilograms of aluminium per one megawatt generated energy, while used aluminium for photovoltaic cells is 45000 kg/MW.

CSP provides over 1000 MW of worldwide electricity, which looks to reach to 15000 MW in near future regarding to new solar projects in US, Spain, China, Morcco and India. Building these solar fields with the mentioned total capacity by aluminium frames requires 1080 million pounds of extruded aluminium [75].

If it can be assumed that the proportion of extruded aluminium that is used in construction of CSP plants remains 34%, the amount of the required aluminium for CSP plants will be approximately 635,000 tons in the next 20years [75].

Today, the amount of energy that is generated by CSP plants is approximately 0.5GW. Predictions reveal that capacity of solar collecting power plants will be 30GW in 2020, 140GW in 2030, and 800GW in 2050, which show a very rapid growth.

Based on these predictions and estimation of average use of aluminium, total amount of used aluminium in CSP plants will be 1.1 and 8 million tons in 2020 and 2030. The average amount of aluminium used in CSP plants in 2050 will be 51 million tons, which has the potential to be double. As a result, approximately 0.3% and 1.9% of annual aluminium production will be used in CSP plants in the decades of 2010-2020 and 2020-2030, respectively. This proportion will be 5.7% in average for the period of 2031-2050 [48].

Today, extruded aluminium used in photovoltaic solar plants is approximately 12% of total amount of aluminium that are used in this kind of solar power plants. If, like what mentioned in future market of CSP plants, it is assumed that this proportion remain constant in future, approximately 1,500,000 tons extruded aluminium will be used for these systems in the next two decades [75].

As mentioned before, these calculations are based on the assumption that the proportion of used aluminium in solar systems will not increase. Considering the growth of aluminium usage in solar systems during the last years, however, clarifies that the solar industries prefer to use extruded aluminium instead of steel frames. Consequently, demands for aluminium related to steel will increase in the course of time.

According to the report of International Energy Agency published in 2010 [47], about 14GW energy are produced by photovoltaic solar system in the world. It also is predicted that the average capacity of this solar system will be 87GW in 2020, 225GW in 225 and 597GW in 2050. Based on this prediction, total amount of aluminium used in photovoltaic solar system will be 3, 7 and 19 million tons in 2020, 2030 and 2050, respectively. Consequently, 0.64% of total annual aluminium production will be used in PV systems in decade 2010-2020, which will reach to 1.21% in decade 2020-2030 and 1.63% in period of 2030-2050.

Temperature is another important factor in efficiency of the photovoltaic solar systems. Researches have revealed that increasing temperature reduces the efficiency of PV solar cells. So, it is important to set a cooling system for PV cells, which provide an excellent opportunity for aluminium to extend its role in solar cells in near future [75].

Another advantage of aluminium over other kinds of materials that encourages many companies in different fields, especially in solar power systems, to use this metal and its alloys is successful and cheap recycling process for aluminium.

Recycling is an important industrial process that can reduce energy usage, air and water pollution and save money. So, recycling should be considered in any kinds of materials that will be employed in industrial or home applications. Recycling is not successful for all materials. Recycling of plastics, for example, is very problematic, but glass can successfully be recycled. Aluminium is the most successful material in recycling. Its recycling is much cost saving process as producing aluminium goods from recycled aluminium cost less than 10% of producing from bauxite ore. Low melting point is the main cause for cost-saving property of aluminium recycling [49]

2.5. Photovoltaic Modules And Corrosion

The performance of solar fields and their power sale revenue are directly dependent on the reflectivity (for solar thermal) and solar panel absorptivity (for solar PV). Unfortunately the mirrors and panels are subjected to exposure to dust moisture and aerosols resulting in a significant drop in system efficiency [76].

Field data has shown that without any external cleaning, reflectivity dropped off with time at a rate of 0.45 percentage points per day and a loss of 16 percentage points in a month (figure 8). Currently the mirrors are cleaned using demineralized water by high pressure spray, deluge type cleaning methods, and fixing of piping spray systems. These methods, however, require demineralized water in the order of 0.23gallons/m^2 of the area [77].

Due to high accumulation rate of dust large quantities of water are require. The washes unfortunately recovered only 1-3% of the loss in reflectivity. Photovoltaic systems use semi-conductor technology in dust and moisture laden region such as the east coast of Saudi Arabia. There is insufficient experience in dust laden and moisture laden region and the problem is very serious. It has been stated that the accumulation of dust on a PV panel causes a loss of 0.5% in PV efficiency. Currently commercially available films have a barrier property of water vapor transmission rate of 10^{-3}g/m^2 per day or one thousand gram square meter per day at 90%. The barrier oxide film therefore develops cracks and pinholes. The life of PV cell is drastically reduced in desert with expended span of humidity and suspended particles (SO_2, NO_X, Cl) [78].

The harsh environment has put a serious demand for fabrication of moisture resistant, corrosion resistant, water and dust repellent surface to prevent PV cells and module. This problem needs three approaches,

a) A dry approach

b) A wet approach

2.5.1. The Dry Approach

This approach is based on surface modification without using water. In certain cases the surface chemistry may be modified by applying low pressure compressed air and installing the system against the direction of winds, and creating conditions which do not allow the dust particles to settle down. Nanostructured surface could be created with ultra-fine grain size which may provide self-cleaning.

2.5.2. The Wet Approach

The wet approach would depend on fabricating hydrophobic films. One good example is provided by Titanium dioxide films for photovoltaic cells derived from a sol-gel process.[79]

Figure 8 describes the set-up of photo electrochemical cells using nano crystalline TiO_2. A sol-gel process is used for spin coating. After being sensitized with a type the films are for transport of electrons in the photo electrochemical cells.

The particles having a size of ~ 100nm. In this process the heating process and application of TiO_2coating by spin coating technique needs to be controlled to produce high efficiency. The demand to produce dust and water repellent coatings is increasing because of its application in MEMS/NEMS and photovoltaic cells. However a considerable progress has been made in producing hydrophobic and dust repellent coating. The best symbol of hydrophobicity is lotus leaf which exhibits water repellent and self-cleaning characteristic. A detail study of lotus leave shows two levels of structures a nano like hair structures with wax covering the surface and micro scale mounds protruding from the surface figure 9[80].

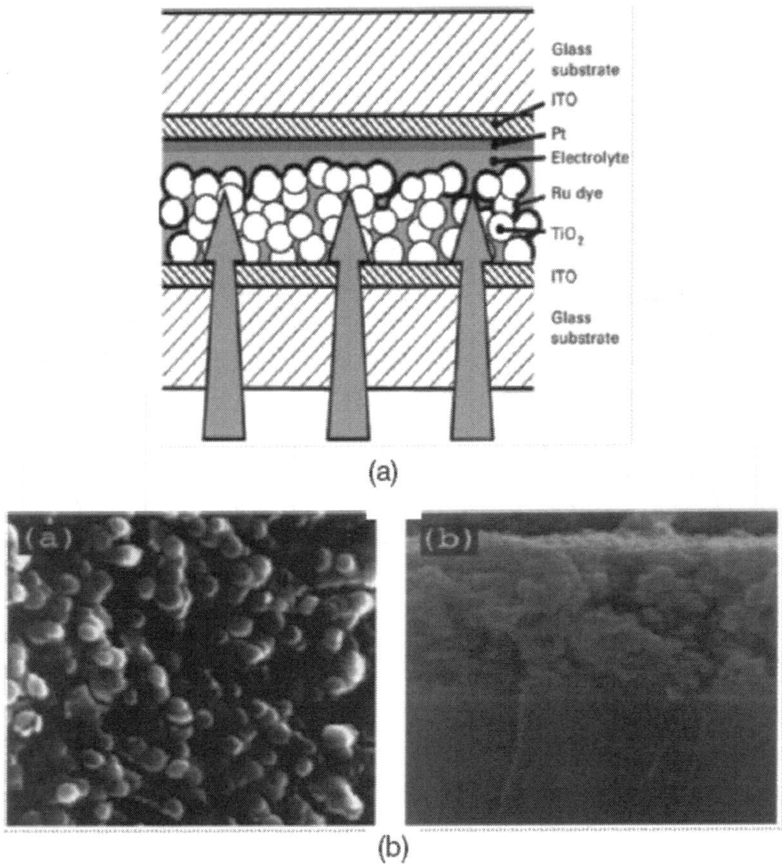

Figure 8. a): Schematic sample setup of the photoelectrochemical cells using nanocrystalline TiO_2. The arrows indicate the direction of incident light [79]. (b): SEM micrographs of a sol-gel TiO2 layer with a thickness of 5 μm on an ITO glass substrate. (a) surface (magnification 15000). (b) cross section (magnification 9000) [79].

Attempts have been made to fabricate nano coatings with inherent self-cleaning properties. The lotus coating is microscopically thin and optically transparent. The coating sheds dust particles utilizing anti- contamination and self-cleaning properties. It is designed to preserve optimum long time performance. The coating can be applied to metal glass surface and epoxy composite substrates. In a desert there may be no rain, however, this technology needs to be complemented by dust management technologies which include, electro dynamic dust shields, thermal control systems and surface power systems. Tribo electric charging has been conducted under Mars surface conditions. The charging showed that large amounts of dust could be accumulated [81].

Figure 9. a) Lotus leaves, which exhibit extraordinary water repellency on their upper side. (b) Scanning eletron microscopy(SEM) image of the upper leaf side prepared by glycerol substitution shows the hierarchical surface structure consisting of papillae,wax clusters and wax tubules. (c) Wax tubules on the upper leaf side. (d) Upper leaf side after critical point (CP) drying. The wax tubules are dissolved, thus the stomata are more visible. Tilt angle 15^0. (e) Leaf underside(CP dried) shows convex cells without stomata [80].

The super hydrophobicity or wettability is characterized by contact angle formed a liquid droplet and solid surface. A surface is consider hydrophobic if the water contact angle (WCA) is greater than 90°and super hydrophobic if the WCA is between 150° and 180°. The contact angle hysteresis is the difference between the advancing angle (ϕ_{adv}) receding (ϕ_{rec}) contact angles. Roughening a surface enhances its water repellent property. The effect of surface roughness has been described by different theories [82].

$$Cos\phi_w = \gamma Cos\phi_y$$

Where ϕ_w and ϕ_y are contact angle on the rough and smooth surface and γ is the roughness factor define as the ratio of solid-liquid area A_{SL} to the projection on a flat plane A_f.

$$\gamma = \frac{A_{SL}}{A_f}$$

According to Wenzel, the space between protrusions is filled by liquid. Cassie and Baxter model [83] assumed that air trapped by asperities of hills on the surface and a composite interface is formed. The composite interface may form air pockets trapped in the cavity. The relationship between apparent contact angles C_β and ideal contact angle is given by

$$Cos C_\beta = r_f\, f Cos\phi_f + (f-1)$$

Where r_f indicates roughness factor of the wetted area and f is the area fraction of the projected wet area. The product r_f is called solid fraction of limit f reached to zero the macroscopic contact angle $C_\beta$approaches to 180° leading to a super hydrophobic behavior. If ϕ_r is high, the drop would be in a Cassie and Baxter and low it would be Wenzel region. Both regions are shown in figure 10 [84].

Two methods are in used for fabrication of super hydrophobic surface; a) making a surface with low energy materials or making a rough surface and coating with low surface materials like tetrafluoroethylene (TEFLCN) or polydimethylsiloxanes (PDMS).

Modifications of a surface with low energy compounds the process such as plasma etching [85] laser etching [86] chemical etching [87] lithography [88] and sol get techniques have been used to produce super hydrophobic surfaces.

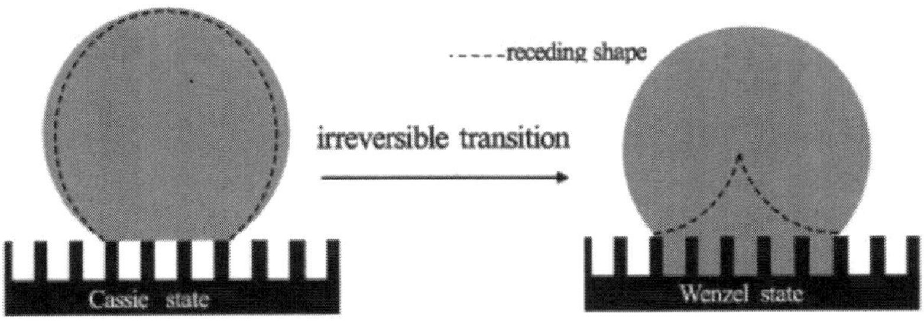

Figure 10. Irreversible Transition from the Cassie State to the Wenzel State by the Wicking Test [84].

Wet coating techniques have been used with success to obtain super hydrophobic surfaces base the critical factor is the preparation of an optimal rough surfaces. Processes such as sand blasting and annealing have been used to obtain a rough surface with grain size in nano dimension [88].

In a study on making a super hydro surface on aluminum samples were mechanically polished to obtain a rough surface in the range of 0.05 to 1.5μm. The surface was modified by treating polydimethyl siloxane vinylterminated (PDMSVT) with 1 weight % Curing agent as cross linked perfluoroethylene (PFPE).

PFPE film was applied by spin coating (3000 rpm). Curing was done for C_9F_{20} at 80°C and PDMSVT and PFPE at 120°. Sol get process titanium coating have been produced and they are extensively reported in literature. Optically transparent super hydrophobic silica based films have been formed by control of hydrolysis and condensation reaction.

In a work under taken [89] three types of originally modified silica gels have prepared as shown in theTable 3.

Table 3. Types of Modified Silica Gel

Sols	Composition	Catalyst	pH	surface bonds after etching
A	TEOS,MPS	HCL	2	Si-O, Si-OH
B	TEOS,MPS	HCL	6	Si-O, Si-OH
C	TEOS	NH_4OH	10	$Si-O_2$, Si-OH

After mixing, the surface chemistry was modified with self assembles monolayers using chlorotrimethylsilane (CTMS) (CH3) (SiCl) and tetra fluorotetrahydroxydimethylchloro silane, hydro phobicity followed the order C, B and A.

Substrate of soda lime glass coated with a thin layer of tin oxide was doped with fluorine (SnO2:T). The transparent conductive oxide (TCO) form, the contact for the solar cell layers deposited on tin oxide. An ethylene vinyl acetate provides both electrical isolation between solar cells and binds the glass back sheet to the module. An extruded aluminum sheet provides structured mounting points and grounding points. Damp heat tests were applied to the modules [90]. A voltage of +600V was applied to the output of one modules and -600V to the other module. The modules work placed in an environmental chamber. Module biased at -600V (amorphous silicon) visible damage occurred. Modules subjected in positive bias showed no or little damaged. The dependence of corrosion damage to polarity suggests the reaction of Na^+ of the glass with fluorine. Humidity acts as an electrolyte for corrosion to take place. Virtually no damage was observed under positive biased polarity. the positive biase only reacts with contaminants outside the glass surface and does not need the mobility of ion.

on subjecting the modules to -600v and 85% RH at 85,72 and 60 °C/ leakage current were observed (fig 11). It was shown that water vapor enhance the leakage of current [89].

Electrochemical corrosion is the result of transport of metal ions between the cell and module frame. under the influence of large cell frame voltage(100-500-V) cell frame voltage present in applications with high system voltages above or below the ground potential [90].The problem is being solved with microencapsulation with low conductivity and control of ionic conductivity of encapsulated free surface and interface. The leakage current is highly dependent on ionic conductivity and moisture level. The transport of ionic elements is associated with corrosion of internal cell material by leakage of current from the cell string to the module exterior. Both processes electro migration and ionic migration need to be controlled [91].

Numerous experiments have been conducted on module, fabricated on the oxide coated glass and operated at voltages and elevated temperatures in a humid climate. The phenomena of electrochemical corrosion appeared to be in all cases related to humidity. Experiments conducted BP Solar a-Si/a-SiGe tendem module fabricated on tin-oxide coated glass and encapsulated with another sheet of glass with EVA, showed that sodium reacted with water to form sodium hydroxide and hydrogen and highly alkaline solution PH>9 such that sodium hydroxide will dissolves Silicate glass rapidly at temperature ($\geq$ 100°C).

In addition, the generation of hydrogen bonds near the interface may lead to a weakening of interfacial bonds because of reduction of tin-oxide. The relative durability of Zinc Oxide appeared to be due to the fact that Zinc oxide is not reduced [92].

A common observation in all investigations shows that corrosion can be minimized by use of low alkali or high resistivity glass, by increasing the adhesion of the transparent conducting oxide to the glass surface or using zinc oxide rather than tin-oxide as a transparent conductive contact. The use of anti-reflection coated (ARC) glass is being used in an increasing percentage of PV modules due to expected high power energy output. The use of ARC glass declined because of the inability of the coating to maintain performance over long period of time. Recent progress made has given some confidence to the consumers to use it again. Several defect such as coating degradation, soiling and optical degradation have been observed. Recent progress in ARC glasses has been shown by ARC glass is developed by the sun power. It has been shown that a well-designed ARC coating protected the glass from humidity and sand blasting, whereas the uncoated glass showed chippings. More than three years of field data showed that the energy gain from ARC significantly exceeded by 3.5 to 5% over uncoated glass which is the consequence of the improved coating gains in diffuse and off angle lights, due to effect of refractive index and light scattering within the coating. The sun power modules are slowly emerging in the market.

Metal electro migration is a big concern in electronic industry along with corrosion, module and discoloration. The metal migration reduces the service life of module. SAFlex PS-41 is the first module to be produced to suppress electro migration by exploring the activity of embedded encapsulation which prevents electro migration when in contact with metals such as silver, copper and nickel. The first encapsulant Saflex PS41 is specifically designed to protect against metal diffusion from solar cell stacks, adhesive and bus ribbons [93].

Whereas enough evidence has been shown how the adverse effects of humidity vapor pressure and current leakages on the glass substrate the back sheet materials in ARC also play a pivotal role. In recent year high moisture barrier and high resistivity coatings on polyethylene terepthalate (PET) have been fabricated for application in PV module back sheet application. It is necessary for the back sheet completely insulating to prevent a conduction path from the back contact to the grounded metal frame. To prevent the penetration of moisture and create low water vapor transmission rates WVTR, g/m^2, d., cost effective coatings have been created on inexpensive polymers such as polyethylene terepthalate (PET) and biaxial oriented polypropylene (BOPP). Table 4 shows the thickness and transmission rates of various coatings applied PET [94].

These back sheets can be used as a substitute for glass if they resist the ingress of moisture and transport of current. The treated polymer has a dramatic improvement over the untreated polymer. The peel strength for different back sheets is shown in Table 5.

Table 4. WVTR for Polymer laminates at 37.8° and 85% Relative humidity

Material	Thickness(mm)	WVTR(g/m²d)
Tedlar/ Al/Tedlar C	0.1	
NREL coated PET	0.18	0.1-0.2
Tedlar / PET/ EVA (TPE)	0.2	3.0
PET	0.1	3.4
EVA	0.4-0.5	27-33

Table 5. Peeling Strengths

Material	t=0	t=1week	t=2weeks
Uncoated PET	1-1.7	—	~0.8
NREL coated PET	7-8	7.10	6.5-6.8
TPE	—	—	~4.0

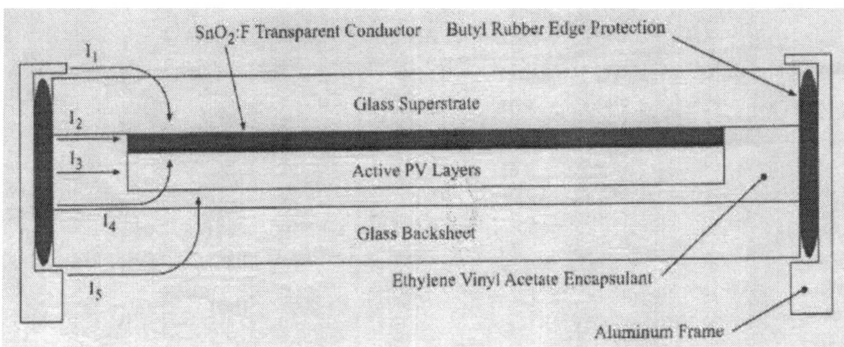

Figure 11. A cross section of a PV module constructed with an SnO_2:F transparent conducting oxide(TCO) layer deposited on a glass superstrate. The active semiconducting layers are deposited over the tin oxide, and the entire package is encapsulated with ethylene vinyl acetate (EVA) between another sheet of glass. Not shown are the laser scribes that form the individual solar cells connected in series. Five possible current paths between the frame and the TCO are illustrated (1) along the surface and through the bulk of the glass superstrate, (2) along the glass superstrate-EVA interface, (3) through the EVA bulk,(4) along the glass backsheet-EVA interface and through the EVA bulk, and (5) along the surface and through the bulk of the glass backsheet, and through the EVA bulk [95].

The above coating shows good adhesion, weather ability and low water vapor transmission rates. These coatings have a good adhesion to EVA after UV or damp heat exposure. The amount of water vapor that diffuses into EVA controls the stability of glass EVA surface. It may that glass/ EVA interface favors condensation reactions and the hydrolysis reactions are difficult to achieve. The exact mechanism is not known, however it is confirmed by research studies that humidity and leakage current play a very predominant role in degradation of PV modules. The important pathways are showing figure 11 [95].

In a study H-V induces leakage current from eight modules, a pair from each module type of C-Si, Pc-Si. (Bulk Si) and tendem junction and multi-junction a-Si were monitored for eight years. It was observed that the leakage currents from C-Si Pc-Si module were thermally activated and the activation energy varied with RH ranging from 0.86-1.0ev at high RH to 0.8ev as low RH. The leakage currents for a-si modules were much lower than bulk Si modules by a factor of 10 to100 times. After operation of HV tests for 24 hrs a day the loss rates were 0.0% year for pL –s, 0.1% year for c-s, (positive polarity) between 0% as 0.05% year for ja-si module. The bulk of the modules degraded compared to modules not biases at HV. For thin film modules the losses were insignificant. The detail of modules discussed as shown in Table 6 [94].

Table 6. Construction and size of PV Modules Tested

Module type	Structure front to back	Area (m^2)	Perimeter (m)	Mounls
C-Si	Glass/ C- Si cells/ Tedar	0 60	3.4	All edges frames
Pe-Si	Glass/ Pc- Si cells/ Tedar	0 52	3.1	All edges frames
2 Ja-Si	Glass/ TCO/a- Si /Al/ glass	0 76	3.7	Rear Brackets
3Ja- Si	Flouro polymer/TCO/a-Si/ SS	0 45	3.3	All edges frames

2.6. Importance Of Dust

Dust is a term which is applied to solid particle with less than 500μm. The main sources of dust are; dust laden winds, volcanic eruptions etc. It also includes micro pollens, microfiber, which are scatter with atmosphere. The factors influence the dust settlements are shown in figure 12 [96].

Dust promotes dust. It settles in region of low vapor pressure induced by the high pressure movements on inclined /vertical surface. The PV system is affected by several environmental factors as shown infigure 13 [96].

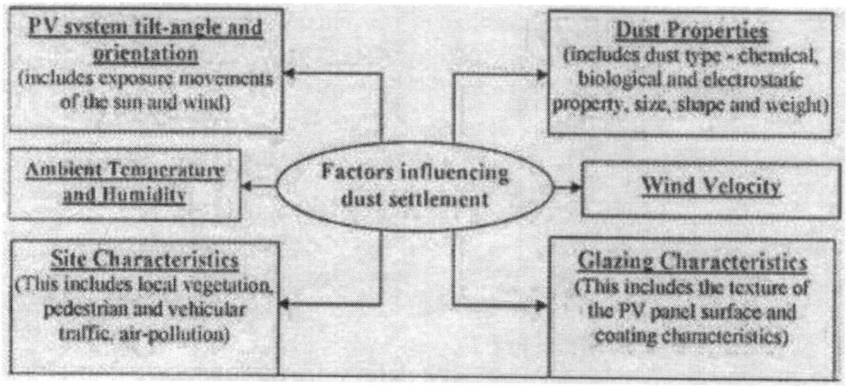

Figure 12. Factors influencing the dust settlement [96].

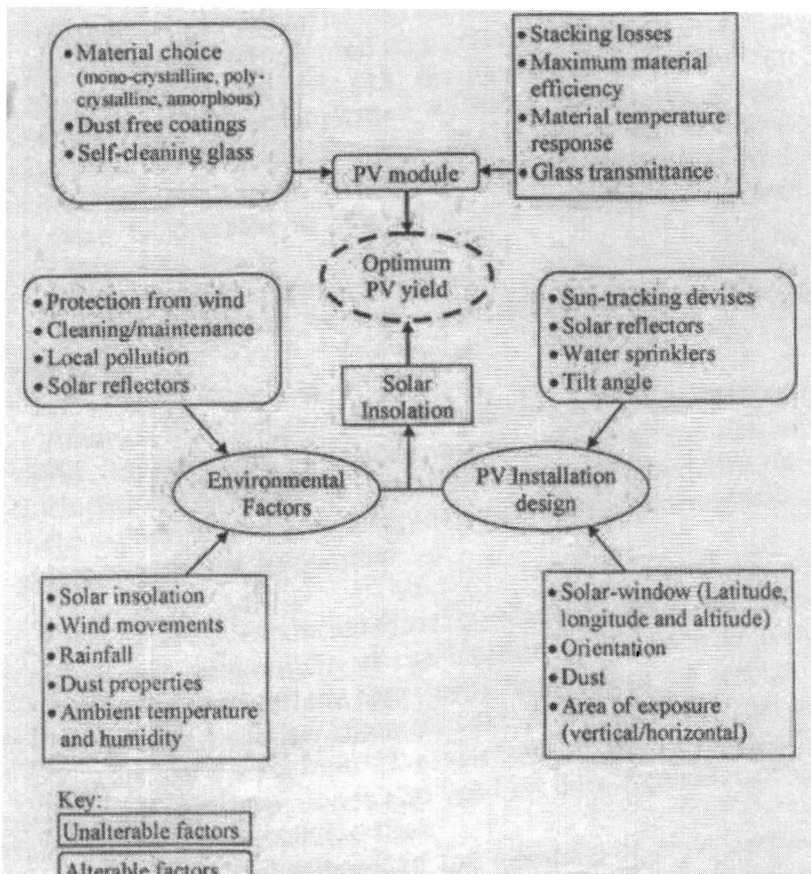

Figure 13. Alterable and unalterable factors determining maximum PV system yield [96].

Amongst all the environmental factors; removal of dust is the most complex factors in PV module. As reported in the literature degradation of 26-40% in the efficiency of thermal panels are photovoltaic cell was reported for installation in Saudi Arabia which is prevalent with dessert wind [97].

The impact of the dust on the reflection of glass in shown figure. Despite few developments in surface technology such as lotus surfaces of work in program with NASA, cleaning by wet method is still predominant.

Future exploration on moon would require mitigating the difficulties posed by Lunor regolith includes dust. NASA is tasked with the development of mitigation strategies of lunar dust. What has been achieved so far is the development of an Electrodynamics dust shield to minimize dust accumulation, a technique which also could be used to remove dust from PVC modules. The dust removal is achieved by applying a multiphase travelling electric field to the electrode that are embedded in the surface to lift and transport charged and uncharged particles off the surface. Following is a brief description of the electrodynamics dust shield technology being developed by NASA. A schematic of three phase Electrodynamics dust shield is shown in figure 14 [98].

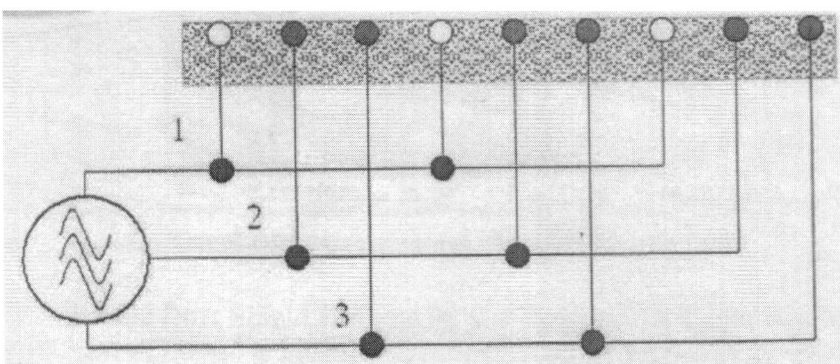

Figure 14. Schematic diagram of a three-phase Electrodynamic Dust Shield [98].

It consists of a series of a parallel electrodes connected to a multiphase AC source that generates a propagating electrodynamics wave. The wave transports the dust particle to a specified location. An electric field is generated by the signal output. The strength of electric field varies proportional to the potential difference between electrodes which is controlled by phase shift. The uniform field carries the charged particle [99].

NASA developed transparent 20cm diameter EDS Indium tin oxide (ITO) electrodes on a polyethylene (PET) film. For testing three electrodes of copper 20cm X 25cm for EDS were constructed and two of them were coated with a lotus film. A lotus coating is a two layer system containing microfills (protrusions) ~2.0 mm, and micro valleys containing epicuticular wax crystals and nano hairs with nano pores. The two layered hierarchical surface is covered by low energy compounds with very low surface energy like PDMS, fluorocarbons and other low energy compounds. Such a two level surface can be created by chemical etching or laser etching to make it a rough surface where the average grain size is in nano region. Sandblasting and short penning or cavitation shotless penning can also be used to make a two level surface. Work on stainless steel has shown the effectiveness of this process. Nanostructured films of TiO_2 were produced by mixing tetra-n-butyltitanate,ethylacetoacetate and ethanol by sol gel technique [100].

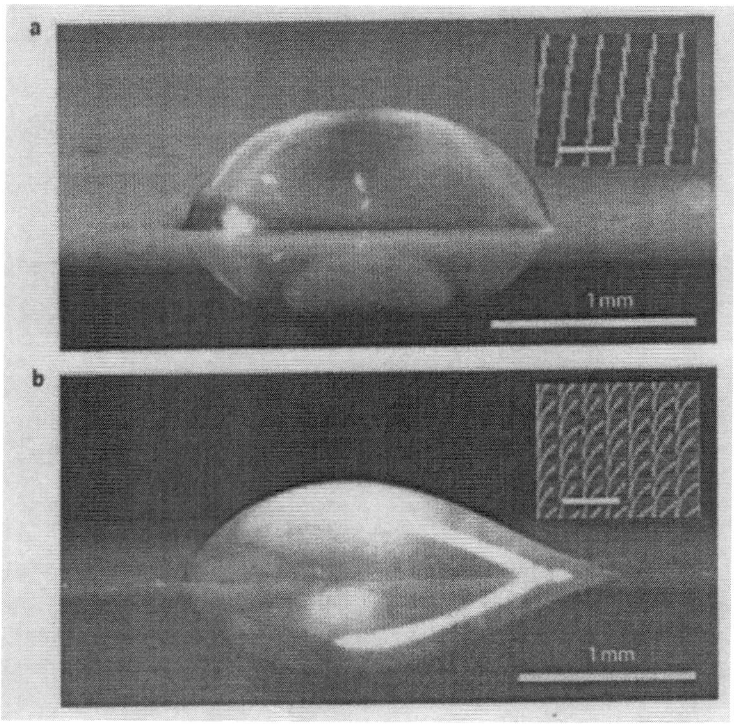

Figure 15. Comparison of wetting behavior on symmetric and asymmetric nanostructured surfaces. a, Axially symmetric liquid spreading of a 1μl droplet of deionized water with 0.002% by volume of surfactants (Triton X-100) deposited on typical vertical nanopillars with diameters of 500 nm, spacings of 3.5 μm and heights of 10 μm(inset).b, Unidirectional liquid spreading of a droplet on the same dimension nanostructures as a, but with a 12^0 deflection angle(inset). The images show the characteristics of a spreading droplet at one instant in time. The scale bars in the insets are 10 μm [102].

The secret lies in preparing a tailored morphology of the surface. The surface can be textured for hydrophobicity (water repelling). The surface exhibit micro convexity with clusters of nanoparticles (99,100). Such surface can trap a large amount of air which has the ability to induce large wet contact angles for hydrophobicity (170^0).Water drops on such surfaces become detached, rolls down and carries the dust with them. However in arid regions, there is hardly any rainfall for this phenomena to occur. Super hydrophobic surfaces can be prepared on metals, glasses and plastics. Similarly a hydrophobic surface can be prepared by depositing films of TiO_2 by hydrolysis of titanium aloxides and hydrolysis of TEOT (Triethylorthotitanate). These films are hydrophilic (water repelling) and also remove contaminants and microbes by the photocatalytic reaction induced by TiO_2 particles. Hydrolysis and condensation of titanium aloxide yield Ti-O based network.

The above surface obtained is hydrophilic. It is also possible to control the direction of flow of water (uni-directional) on asymmetric nanostructured surface by allowing the liquid to flow in one direction and pin on other direction which can be very helpful for various geometries of photovoltaic modules [101]. Figure 15 shows a comparison of wetting behavior on symmetric and asymmetric nanostructure surface [102].

The lotus coatings being worked out by NASA is developed on the principle described above.

The lotus surface developed is expected to mitigate dust without use of water as in the case of hydrophobic surface. The lotus coating two level would shed particles utilizing anti contamination and self cleaning properties which would minimize dust accumulation. Such coatings based on the structure of lotus flower such as hydrophobic coatings on glass and plastics would have the capability to repell dust. NASA is developing both hydrophobic and hydrophilic coatings which are next generation coatings to minimize dust on solar cells and thermal radiation. Self cleaning and anti-contamination systems being developed have also the capability to kill bacteria, chemical agents, pathogens and environmental pollutants. In future the super hydrophobic coatings would play a leading role not only in lunar environment but also in solar cells and most importantly in space exploration. The hybrid coating for photovoltaic solar arrays are shown in figure 16 [103].

In order to understand the working of EDS, it is important to understand the forces which are responsible for lifting the sands.Two types of forces are applied by the electrodynamic field; a: Electrostatic force and b): di electrophorelectric force.

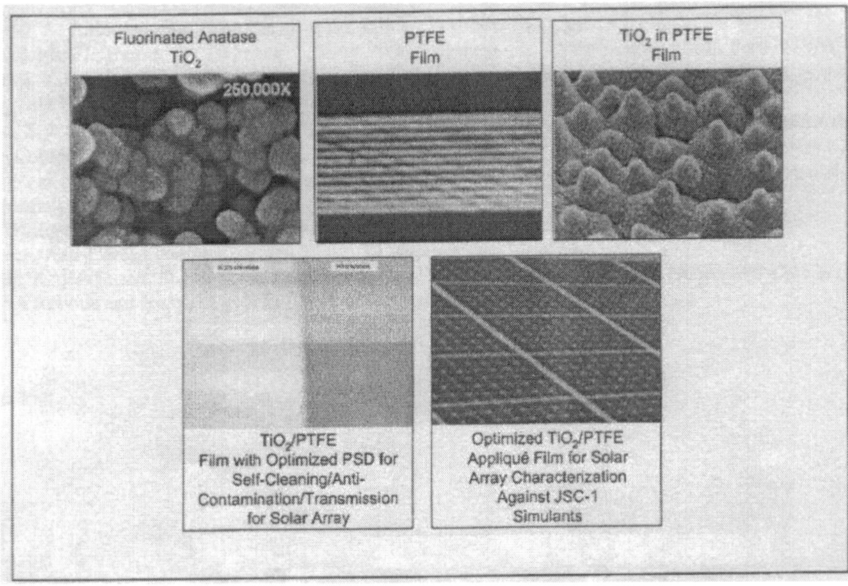

Figure 16. Hybrid coating for Photovoltaic solar arrays [103]

Most airborne dust particles acquire an electrostatic charge during their detachment process. Each sand particle is subjected to a sinusoidal excitation voltage generated by the electric field. A charge particle experienced two forces of repulsion, one tangential and other normal to the contact angle. The lift force for the particle is provided by centrifugal force which is induced by the curvilinear motion of the particle. Another particle charged with –q will be subjected to repulsive force and it would levitate of the lifting force is larger than the adhesion force due to the cumulative effects of lifshitz – vander walls forces, electrostatic forces and capillary forces. There is hardly any capillary force in the desert region. On energizing of three phase voltage,the charged particles are lifted from the surface by the vertical component of the field and the travelling wave component as mentioned earlier carries the dust to the screen. Single phase excitation lifts the particles. This process becomes more effective when a three phase voltage is applied.

Another force to be considered is dielectrophoretic force. It is experienced by charged or uncharged particles in any AC or DC field (E). Because the particles +q and – q are charged and separated by a distance, a dipole moment (qd) is formed. Because of induced dipole moment these particles experience dielectrophoretic force. The applied voltage creates a gradient in the electric field. The divergence of electric field applies a dielectrophoretic force F_d and a torque T. This force causes the movement of neutral particles on the surface and induces electrostatic charging by triboelectrification. This acquired charge would induce to the columbic force of repulsion to lift the particles. In summary, columbic and dielectrophoretic forces move the dust particles to the surface and hence the particles acquire charge. These charge particles are repelled by electrostatic forces.

This mechanism applies also to conducting particles deposited on the shield. The charge q is proportional to E^2. Particle is the vicinity of electrodes acquire electrostatic charge and they are repelled when the force of repulsion $F_{Repulsion} = 9 E_0 = E_0 r^2$ is > force of adhesion ($F_{Adhesion}$).The particle would be lifted.

By applying a three phase voltage 90% of the dust is removed in about two minutes.

The energy requires for dust removal is only a small fracture of the energy output of the modulesc [104].

The electrode grid uses indium tin oxide or carbon nanotubes. Figure 11 shows transparent EDS coatings in glass.The role of aluminium has become very predominant in solar power system. The solar power system has been divided in four distinct groups, parabolic trough, parabolic dish, linear fersenel and solar tower. Aluminium is one of the most important material utilized in solar absorbing due to the capability of its anodic layer formed on its surface by the process of anodization. Eutectic binary aluminiumm alloys such as Al-0 wt% Ni, Al-33 wt% Cu and Al-7.5wt% Ca have been successfully used as absorber(low reflection and high absorption).The mechanical and thermal ability of aluminium alloys and regeneration of surface is etching enhances their properties in solar power system.

Aluminium extrusion provides a clear economic advantage in the product of solar application. White steel costs less than aluminum on a dollar per pound basis, the lower weight of aluminium (1/3rd of steel) allow far more material to be used at a lower cost.

Because of its recyclability,light weight, high strength and high corrosion resistance,it has become a preferred material. By using aluminium Alcoa is saving on costs of solar pond and transportation.

Hydro in Germany is making mirrors for concentrated solar power as well as absorber sheets for solar thermal application. This company launches the first tailored Hybridlife solar aluminium alloy for concentrated solar power and launched high select coatings for solar thermal systems. Hydro serves its customer in solar thermal concentrated solar power and photovoltaic for all area of solar energy product.

Ali Baba has manufactured 1 KW, 2 KW, 3 KW and 10 KW solar power systems, aluminium based with a life span of 25 years.(AliBaba.com). Light technology is a leading and provides standard aluminium profile for mounting systems for solar energy.

Pacific power management (PPM) has announced the installation of a 800 KW power plant for Sierra Aluminium Company. It contains 4,480 Mistzubish Electric,180 watts modules and 2 satcons 5oo Kw invertors. It generates 1.4m Kwh per year.

It would reduce Sierra carbon footprint by 48% and supply power to 28,000 homes over a 25 year period.

The use of solar mirrors could reduce the cost by 20%. For a 50 megawatt power plant, a saving of 20 million euros could be made.

3. CONCLUSION

Aluminum is playing a predominant role in solar power system because of its technical capability,ease of fabrication and ease of transport use, recyclability and resistant to corrosion. The promising future of aluminium in solar power is reflected by the projections on market growth from 210 mm^2 to 11 bmm^2. By 2050, the amount could reach 39 mtons from the existing 17 mtons. The major attributes are large energy area for collection, solar directed installation and dynamic development. However there are several technical problems associated with solar power such as the ingress of moisture causing corrosion and leakage of current causing deterioration of modules. The water vapours ingresses through the edges and increases the conductivity of the front glass surface and also the magnitude of leaking current. In the four types of modules a). C-Si, b). PC-Si, c). 2J a-Si (Glass/TCO/a-Si/Al/Glass) and 3 Ja-Si (Fluropolymer/TCO/a-Si/stainless steel), the two modules containing a-Si showed the maximum resistance to HV operation. In HV operation, all modules degrade at rates higher than the modules not biased on HV. Films on PET showed promising properties as a back sheet replacement for glass. These coatings exhibit excellent moisture resistance properties and a good cohesion after exposure to damp heat. The corrosion effect can be minimized by increasing adhesion of transparent oxide by using Zinc oxide in place of Tin oxide and by using low acetate and high resistivity glass.

Dust is still haunting the scientists and engineers working on solar and space equipment. It is of vital importance to solar panels and equipment used in space exploration. A substantial amount of research has been done on electrodynamics system to remove dust. This is couples with creating a lotus surface (two level) hierarchical surface (nano/micro hybrid) to create self cleaning properties for removal of dust by mimicking the surface of a lotus flower. Various paints containing self cleaning agents have also been designed to remove dust. The wet chemistry route creating a superhydrophobic surface is an outstanding achievement but it cannot be applied in dessert conditions. Intensive work is undertaken by NASA to create dust shields. It appear that new techniques would be developed to mitigate the degradation of PV modules and the use of aluminum would continue to rise.

NAN. ACKNOWLEDGEMENTS

I am highly indebted to Ms. Zahra Khan (Comsats) and Ms. Tayyaba Abid (Comsats) for their dedicated help in preparing the manuscript of the chapter for the book. The above colleagues have put in a very dedicated work in all technical aspects related to the formatting processing and editing of this script.

REFERENCES

1. T.K. Ghosh MAP. Energy resource and systems, Volume 2: Renewable sources. 1st ed. : Springer; 2011.
2. Xiaoming L., Duowang F., Fan Y. A study of the organization and performance of thermally evaporated aluminium reflector for solar energy system. 2nd Conference on Environmental Science and Information Application Technology, ESIAT 2010, July 17, 2010 - July 18; 2010; Wuhan, China: IEEE Computer Society; 2010.
3. Mazria E. Passive solar energy book. Emmaus: Rodale Press; 1979.
4. Stine WB HR. Solar energy fundamentals and design. New York: Wiley; 1985.
5. Kreith F KJ. Principles of solar energy. Washington, DC: Hemisphere Publishing; 1978.
6. Tiwari G,. Solar energy: fundamentals, design, modeling and applications. New York: CRC; 2002.
7. Gupta PK. Renewable energy sources-a longway to go in India. Renewable Energy 1999;16(1-4):1216-9.
8. Ummadisingu A, Soni MS. Concentrating solar power - Technology, potential and policy in India. Renewable and Sustainable Energy Reviews 2011;15(9):5169-75.
9. Anderson B. Solar energy: fundamentals in building design. New York: McGraw-Hill; 1977.

10. Malik MAS, Tiwari GN, Kumar A, Sodha MS. Solar distillation. New York: Pergamon Press; 1985.

11. Kreider JF KF. Solar heating and cooling. New York: McGraw-Hill; 1977.

12. Kalogirou S. Solar water heating in Cyprus: current status of technology and problems. Renewable Energy 1997;10(1):107-12.

13. Kalogirou SA. Solar thermal collectors and applications. Progress in Energy and Combustion Science 2004;30(3):231-95.

14. Bean JR., Diver RB. Performance of the CPG 7.5-KWe dish-stirling system. Proceedings of the 28th Intersociety Energy Conversion Engineering Conference, August 8, 1993 - August 13; 1993; Atlanta, GA, USA: Publ by SAE; 1993.

15. Corrigan RD, Peterson TT, Ehresman DT. Update of the Solar Concentrator Advanced Development Project. Part 1 (of 6): Aerospace Power Systems and Power Conditioning, August 6, 1989 - August 11; 1989; Washington, DC, USA: Publ by IEEE; 1989.

16. Dang A. Concentrators: A Review. Energy Conversion and Management 1986;26(1):11-26.

17. Fend T, Hoffschmidt B, Jorgensen G, Kuster H, Kruger D, Pitz-Paal R, et al. Comparative assessment of solar concentrator materials. Solar Energy 2003;74(2):149-55.

18. Zhao J, Wang A, Blakers AW, Green MA. High efficiency prismatic cover silicon concentrator solar cells. Twentieth IEEE Photovoltaic Specialists Conference - 1988, September 26, 1988 - September 30; 1988; Las Vegas, NV, USA: Publ by IEEE; 1988.

19. Gregory GG, Koshel RJ. Modeling the operating conditions of solar concentrator systems. Photonics for Solar Energy Systems, April 5, 2006 - April 6; 2006; Strasbourg, France: SPIE; 2006.

20. Hasuike H, Yoshizawa Y, Suzuki A, Tamaura Y. Study on design of molten salt solar receivers for beam-down solar concentrator. Solar Energy 2006;80(10):1255-62.

21. Heath J, A.R., Hoffman EL. Recent gains in solar concentrator technology. J.Spacecraft Rockets 1967;4(5):621-4.

22. Hinsch A, Zastrow A, Wittwer V. Sol-gel glasses. A new material for solar fluorescent planar concentrators? Solar energy materials 1990;21(2-3):151-64.

23. Irshid MI, Othman MO. V-Troughs with High Concentration Ratios for Photovoltaic Concentrator Cells. Solar Cells 1988;23(3-4):159-72.

24. Isshiki N, Watanabe H, Shishido K, Ohtomo M, Watanabe K. Studies on solar-dish heated stirling engines TNT-3, NAS-2. Proceedings of the 28th Intersociety Energy Conversion Engineering Conference, August 8, 1993 - August 13; 1993; Atlanta, GA, USA: Publ by SAE; 1993.

25. Kribus A, Huleihil M, Timinger A, Ben-Mair R. Performance of a rectangular secondary concentrator with an asymmetric heliostat field. Solar Energy 2000;69(2):139-51.

26. Li M, Wang LL. Investigation of evacuated tube heated by solar trough concentrating system. Energy Conversion and Management 2006;47(20):3591-601.

27. Maish AB. PV Concentrator Array Field Performance Measurement.Solar Cells 1986;18(3-4):363-71.

28. Morel DE, Ayers SR, Gulino DA, Tennyson RC, Egger RA. Solar Concentartor Materials Development. 21st Intersociety Energy Conversion Engineering Conference: Advancing toward Technology Breakout in Energy Conversion. San Diego, CA, USA: ACS; 1986.

29. Palavras I, Bakos GC. Development of a low-cost dish solar concentrator and its application in zeolite desorption. Renewable Energy 2006;31(15):2422-31.

30. Rowan B, Mc Cormack S, Doran J, Norton B. Quantum dot solar concentrators: An investigation of various geometries. High and Low Concentration for Solar Electric Applications II, August 26, 2007 - August 28; 2007; San Diego, CA, United states: SPIE; 2007.

31. Singh P, Liburdy JA. Solar concentrator design for uniform flux on a flat receiver. Energy Conversion and Management 1993;34(7):533-43.

32. Tecpoyotl-Torres M, Campos-Alvarez J, Tellez-Alanis F, Escobedo-Alatorre J, Aguilar JQ, Sanchez-Mondragon J. RF control system of a parabolic solar concentrator. High and Low Concentration for Solar Electric Applications II, August 26, 2007 - August 28; 2007; San Diego, CA, United states: SPIE; 2007.

33. Valade FH. Solar concentrator advanced development program update. Proceedings of the 23rd Intersociety Energy Conversion Engineering Conference, July 31, 1988 - August 5; 1988; Denver, CO, USA: Publ by IEEE; 1989.

34. Wong WA, Geng SM, Castle CH, Macosko RP. Design, fabrication and test of a high efficiency refractive secondary concentrator for solar applications. 35th Intesociety Energy Conversion Engineering Conference, July 24, 2000 - July 28; 2000; Las Vegas, NA, USA: IEEE; 2000.

35. Holbert KE, Haverkamp CJ. Impact of solar thermal power plants on water resources and electricity costs in the Southwest. 41st North American Power Symposium, NAPS 2009, October 4, 2009 - October 6; 2009; Starkville, MS, United states: IEEE Computer Society; 2009.

36. Ortiz-Rivera E, Feliciano-Cruz L. Performance evaluation and simulation of a Solar Thermal Power Plant. 2009 IEEE Energy Conversion Congress and Exposition, ECCE 2009, September 20, 2009 - September 24; 2009; San Jose, CA, United states: IEEE Computer Society; 2009.

37. Cambronero LEG, Canadas I, Martinez D, Ruiz-Roman J. Foaming of aluminium-silicon alloy using concentrated solar energy. Solar Energy 2010;84(6):879-87.

38. Flamant G, Ferriere A, Laplaze D, Monty C. Solar processing of materials: opportunities and new frontiers. Solar Energy 1999;66(2):117-32.

39. Martinez D, Rodriguez J. Materials surface treatments by concentrated solar light: A renewable energy option; Tratamiento superficial de materiales mediante luz solar concentrada: una opcion mediante energias renovables. Revista de Metalurgia (Madrid) 1998;34(2):104-8.

40. Bakos J, Miyamoto HK. Solar hydrogen production - Renewable hydrogen production by dry fuel reforming. Solar Hydrogen and Nanotechnology, August 14, 2006 - August 17; 2006; San Diego, CA, United states: SPIE; 2006.

41. Baykara SZ. Hydrogen production by direct solar thermal decomposition of water, possibilities for improvement of process efficiency. Int J Hydrogen Energy 2004;29(14):1451-8.

42. Mori M, Kagawa H, Nagayama H, Saito Y. Current status of study on hydrogen production with space solar power systems (SSPS). 4th

International Conference on Solar Power from Space, SPS '04 - Together with The 5th International Conference on Wireless Power Transmission, WPT 5, June 30, 2004 - July 2; 2004; Granada, Spain: European Space Agency; 2004.

43. Roman R, Canadas I, Rodriguez J, Hernandez MT, Gonzalez M. Solar sintering of alumina ceramics: Microstructural development. Solar Energy 2008;82(10):893-902.

44. Brenna M, Foiadelli F, Roscia M, Zaninelli D. Evaluation of solar collector plant to contribute climate change mitigation. 2008 IEEE International Conference on Sustainable Energy Technologies, ICSET 2008, November 24, 2008 - November 27; 2008; Singapore, Singapore: Inst. of Elec. and Elec. Eng. Computer Society; 2008.

45. Ronnelid M, Adsten M, Lindstrom T, Nostell P, Wackelgard E. Optical scattering from rough-rolled aluminium surfaces. Appl.Opt. 2001;40(13):2148-58.

46. Machinda GT, Chowdhury S, Arscott R, Chowdhury SP, Kibaara S. Concentrating solar thermal power technologies: A review. 2011 Annual IEEE India Conference: Engineering Sustainable Solutions, INDICON-2011, December 16, 2011 - December 18; 2011; Hyderabad, India: IEEE Computer Society; 2011.

47. IEA. Technology roadmap – concentrating solar power. 2010;.

48. Bödeker JM, Bauer M, Pehnt M. Aluminium and Renewable Energy Systems – Prospects for the Sustainable Generation of Electricity and Heat. 2010;.

49. Mok SC. Aluminium economy for sustainable development: Aluminium as core material for energy storage and energy saving products: Low cost, high performance, and easy processing in developing countries. 2011 IEEE Global Humanitarian Technology Conference, GHTC 2011, October 30, 2011 - November 1; 2011; Seattle, WA, United states: IEEE Computer Society; 2011.

50. Fend T, Jorgensen G, Kuester H. Applicability of highly reflective aluminium coil for solar concentrators. Solar Energy 2000;68(4):361-70.

51. Granqvist CG. Radiative Heating and Cooling with Spectrally Selective Surfaces. Appl.Opt. 1981;20(15):2606-15.

52. Keller F, Hunter MS, Robinson DL. Structural features of oxide coatings on aluminium. Electrochemical Society -- Journal 1953;100(9):411-9.

53. Henley V. Anodic Oxidation of Aluminium and its Alloys. : Pergamon Press; 1982.

54. Woodman TP. Light Scattering in Porous Anodic Aluminium Oxide Films-1. Colour Effects. Thin Solid Films 1972;9(2):195-206.

55. Woodman TP. Light Scattering in Porous Anodic Aluminium Oxide Films - 2. Polarization Effects. Thin Films 1972;9(3):389-94.

56. Pavlovic T, Ignatiev A. Optical And Microstructural Properties of Anodically Oxidized Aluminium. Thin Solid Films 1986;138(1):97-109.

57. Pavlovic T, Ignatiev A. Optical Properties of Spectrally-Selective, Anodically-Coated, Electrolytically-Colored Aluminium Surfaces. Solar energy materials 1987;16(4):319-31.

58. Granqvist C. Optical properties of integrally colored anodic oxide films on aluminium. Journal of Applied Physics 1980;51(6):3359-61.

59. Kumar SN, Malhotra LK, Chopra KL. Nickel Pigmented Anodized Aluminium as Solar Selective Absorbers. Solar energy materials 1983;7(4):439-52.

60. Cody GD, Stephens RB. Optical Properties of a Microscopically Textured Surface. 1978;40:225-39.

61. Chang V, Bolsaitis P. Study of Two Binary Eutectic Aluminium Alloys as Selective Absorbers for Soalr Photothermal Conversion. Solar energy materials 1980;4(1):89-100.

62. Pellegrini G, Brughera P, Quazzo F. On the Properties of the Superplastic Aluminium-Calcium Alloy as Material for Solar Collectors. Solar energy materials 1982;7(3):351-7.

63. Parida B, Iniyan S, Goic R. A review of solar photovoltaic technologies. Renewable and Sustainable Energy Reviews 2011;15(3):1625-36.

64. Wurfel P. Thermodynamic limitations to solar energy conversion. : Elsevier; 2002.

65. Feltrin A, Freundlich A. Material considerations for terawatt level deployment of photovoltaics. Renewable Energy 2008;33(2):180-5.

66. Duran Sahin A, Dincer I, Rosen MA. Thermodynamic analysis of solar photovoltaic cell systems. Solar Energy Mater.Solar Cells 2007;91(2-3):153-9.

67. Brown GF, Wu J. Third generation photovoltaics. Laser and Photonics Reviews 2009;3(4):394-405.

68. Miyashita N, Shimizu Y, Kobayashi N, Okada Y, Yamaguchi M. Fabrication of GaInNAs-based solar cells for application to multi-junction tandem solar cells. 2006 IEEE 4th World Conference on Photovoltaic Energy Conversion, WCPEC-4, May 7, 2006 - May 12; 2006; Waikoloa, HI, United states: Inst. of Elec. and Elec. Eng. Computer Society; 2007.

69. Green MA. Thin-film solar cells: Review of materials, technologies and commercial status. Proceedings of the International Conference on Optical and Optoelectronic Properties of Materials and Applications (ICOOPMA 2006) 233 Springer Street, New York, 10013-1578, United States: Springer New York LLC; 2007.

70. Rey-Stolle I, Garcia I, Galiana B, Algora C. Improvements in the MOVPE growth of multi-junction solar cells for very high concentration. J.Cryst.Growth 2007;298:762-6.

71. Conibeer G, Li Y, Slaoui A, Tao M, Topic M. Advanced inorganic materials and concepts for photovoltaics. Energy Procedia 2011;10:v-.

72. Bosi M, Pelosi C. The potential of III-V semiconductors as terrestrial photovoltaic devices. Prog Photovoltaics Res Appl 2007;15(1):51-68.

73. Jager-Waldau A. European Photovoltaics in world wide comparison. J.Non Cryst.Solids 2006;352(9-20):1922-7.

74. Tjotta S. Advantages of extruded aluminium in solar power generation systems processes. Advanced Materials and Processes 2011;169(1):28-9.

75. Werner C. Aluminium extrusion in solar power applications. Light metal Age 2010(August):24-8.

76. G.E. Cohen, D.W. Keatney and G.J. Kolb, Final report on "Operation and Maintenance Improvement Program for Concentrating Solar Power Plants" Sandia lab report CSPSAND,1998.

77. Parabolic Trough Workshop "Cooling for Parabola Trough Power Plants, DOE presentation, NREL, PR-550-4002r,August 16-18,1999, Ontrio 2007.

78. Zaki Ahmad and M. Ahsan, Nano-paints for Desert Environment, jr of xyz or ACE, Nov 2006

79. Yongxiang Li, Jurgen Hagen, Winfried Schaffrath, Peter Ostrchik, Dielet Harry Solar Energy Materials and Solar Cells, 56(1995)[67-67]

80. Hans J, Ensikat, Ditsche-Kuru P, Neinhuis C, Barthlott W. Superhydrophobicity in perfection: the outstanding properties of the lotus leaf. Beilstein Journal of Nanotechnology 2011; 2, 152–161.

81. Mark J. Hyatt, Sharin A. Starka, Dust Management Project Advance Capabilities project office, 21000 brookport Rd, Glenn research Centre at lewis fields, Cleveland, OH 44135 USA,2006.

82. R.N Wenz, Resistance to solid surface to wetting by wear, Ind Eng. Chem vol 28, p 546, 1944

83. Yongxiang Li, Jurgen Hage, Winfried Schaffartg,Peter OOtschik,Dieter, Harer, Titanium Dioxide Films For Photovoltaic Cells Derived From Sol Gel Processes,Solar Energy Materials And Soalr Cells,56,1991,page no 166-174.

84. Wu XD, Zheng LJ, Wu D. Fabrication of superhydrophobic surfaces from microstructured ZnO-based surfaces via a wet-chemical route. Langmuir 2005; 21, 2665-2667

85. M.H. Jin, X.J Feng, J.M. Xi, J. Zhai, K.W. Chow, Feng etal; Super hydrophobic PDMS surface with low Adhesive Forces, Micro molecular rapid communication, vol. 26,2002 pp1805.

86. X.Y. Song, J. Zhai, Y.L. Wang, L. Jiang; fabrication of super hydrophobic surface by self-assembly and their water adhesion property, J. Phy chem 13, vol. 109, 2005 pp4048-900.

87. Z. Gaou, F.Zhau, J.Hao, W.Liu, stable Biomimetic superhydrophobic Engineering Materials J. Am. Chem. Soc,Vol 127,2005pp 15670-15070.

88. Y.H.Yang, Z.Y.Lee, B.Wang. C.X.Wan, D.H.Chen,K.G.N.Yang, Mehanical ,Electochem and Tribological Properties of Nanocrystalline surfaces,306 stainless steel,J Phy.condens,mats, Vol 17 ,2005 pp 5441-5446.

89. G.Z Cao, H. Tian ,Synthesis of highly porous inorganic Hybrids by Ambient Pressure, sol gel processing, Jr Sol Gel.Tech, Volume 13,1998p 315.

90. C.R.Osterwall, T.J.Cueto, Electro chemical Corrosion of SnO2 F. Transparent conducting layers in thin film photovoltaic module, Solar energy, Materials and solar cells,79(2000),21-33(2003).

91. RG Ross,Jr, G.R.Mon,L.C.Wen and R.S sigma, Measurement and characterization of voltage and current induced Degradation of thin film Photovoltaic modules, Solar Cells,27(1989) pp 289-298.

92. Kumar S, Drevillon, A Real time Eilipsometry study of growth of Amorphous silica on transparent conducting oxide, Jr. App Phy 1989,65(8),pp 3023-3034.

93. Solution Inc, 575, Maryvillee Centre Drive St Louis, Missouri,6314 USA,2012.

94. G.D.Barbar,G.J.Jorgensen, K.Terwilliger,S.H.Glick.J.Pern, T.J McMohan,29TH IEE Specialist Conference , New Orlenas, Louisana, May 20-24,2002.

95. Osterwald CR, McMahon TJ, del Cueto JA. Electrochemical corrosion of SnO2:F transparent conducting layers in thin-film photovoltaic modules. Solar Energy Materials and Solar Cells 2010; 79, 21–33.

96. Mani M, Pillai R. Impact of dust on solar photovoltaic (PV) performance: Research status, challenges and recommendations. Renewable and Sustainable Energy Reviews 2010; 14, 3124–3131.

97. Nimmo B, Seid Sam, Effect of dust on performance of Thermal and Photovoltaic plate collector in Saudia Arabia in Veziroglu, TN, Ed, Proc 2nd Miami Conf,1979,Page 33-35.

98. Calle CI, Immer CD, Ferreira J, Hogue MD, Chen A, Csonka MW, Van Suetendael N, Snyder SJ. Integration of the Electrodynamic Dust Shield on a Lunar Habitat Demonstration Unit. ESA Annual Meeting on Electrostatics 2010; Paper D1.

99. C.I.Calle, A.Chen, C.D.Immer, M.Csonoka, M.D.Hogue,S.J.Synder, M.Rodriguef and D.R.Margretta, ARSC.Aerospace Kennedy space Centre,3289. Dust removal Technology Demonstration for a lunar Habital,2010.

100. L.Hu, T.Yoko,H.Kozuka, S.Sakka, Jr Thin solid Films(1992), 18,P 219.

101. Kuang.Haczu, Rongxia and Evelyn. Wang, Unidirectional Liquid spreading on asymmetric nanostructure surface, Nanomaterial Letters,28th March,2010, DOI:10-10.38/NM7276.

102. Chu KH, Xiao R, Wang EN. Uni-directional liquid spreading on asymmetric nanostructured surfaces. NATURE MATERIALS 2010; 9(5), 413-417.

103. Pirich R, Weir J, Leyble D. Self-Cleaning and Anti-Contamination Coatings for Space Exploration: An Overview. Conference on Optical System Contamination - Effects, Measurements, and Control 2008: Proceedings of the Society of Photo-optical Instrumentation ENGINEERS (SPIE), AUG 13-14, 2008, San Diego, CA; 2008.

104. Malay Mazumdar, Adhesion society Meeting,Feb,14-16,2011, Boston, MA 002215.

Index

1xxx Series Alloys, 4
2xxx Series Alloys, 4
3xxx Series Alloys, 4
5xxx Series Alloys, 5
6xxx Series Alloys, 5
7075 Al alloy, 61, 188, 189, 190, 195, 202, 204
7xxx Series Alloys, 6
8xxx Series Alloys, 6

A

Aluminium, 1, 2, 4, 7, 10, 12, 48, 49, 53, 58, 61, 63, 65, 72, 74, 78, 88, 96, 108, 109, 110, 111, 112, 113, 115, 124, 169, 223, 225, 228, 229, 230, 235, 237, 252, 253, 258, 259, 260
anodization, 116, 152, 153, 155, 161, 162, 163, 164, 167, 252
anodized aluminum oxide, 116
artificial aging, 54, 127, 128, 131, 132, 134, 144, 153, 158, 159, 168

C

Copper, 4, 15, 42, 44, 108, 111, 153
corrosion, 1, 2, 3, 4, 5, 6, 7, 8, 9, 10, 12, 13, 14, 15, 16, 17, 18, 19, 20, 21, 22, 23, 24, 26, 27, 168, 169, 224, 235, 238, 243, 244, 252, 253, 262
Creep Ductility, 101, 111
cryoforging, 151, 152, 153, 165

D

Deep Rolling, 169, 172, 175, 176
Dust, 246, 248, 253, 261, 262

E

Ecap, 65, 68, 70, 72
electrochemical impedance spectroscopy, 11, 28, 35, 38, 48, 51, 115, 116, 117, 121
Evacuated Tube Collectors, 230

F

Flat-Plate Collectors, 230

G

gas metal arc welding, 128
Grain Boundary Sliding, 91, 112

H

Hardness, 69, 175, 177

I

Iron, 17, 53

L

light poles, 128, 129, 130, 147, 148
Linear Fresnel, 227, 228
Lithium, 6, 18, 42, 46, 51, 52

M

Magnesium, 5, 14, 109, 167
Manganese, 4, 17, 57
micro-nanostructure, 116, 119
microstructure, 3, 5, 10, 12, 14, 20, 22, 24, 27, 39, 41, 45, 48, 55, 59, 62, 64, 66, 72, 74, 75, 208, 211, 212, 214, 216, 217, 219, 220, 221, 232

P

Parabolic Dish, 226
Parabolic Trough, 226, 261
phase transformation, 188, 201
Photovltaic Solar System, 233
post weld heat treatment, 127, 128, 144, 147, 148, 149, 150
potentiodynamic polarization, 116, 117, 121

R

resistance, 1, 2, 4, 5, 7, 10, 11, 14, 16, 26, 30, 36, 50, 51, 54, 58, 60, 96, 97, 100, 105, 106, 115, 116, 117, 121, 122, 123, 124, 125, 126, 128, 151, 152, 153, 155, 163, 165, 168, 169, 187, 190, 211, 214, 217, 235, 252, 253

S

semi-solid, 187, 188, 189, 190, 191, 192, 193, 196, 197, 198, 200, 202, 203, 208, 209, 211, 212, 214, 216, 218, 219, 220, 221
Silicon, 5, 15
SIMA, 187, 188, 190, 192, 193, 197, 198, 199, 203, 205, 212, 213, 214, 215, 216, 217, 218, 219, 220
Solar Power, 223, 225, 235, 258, 260
Solar Thermal Collectors, 229
superhydrophobic aluminum, 116, 125

T

Tensile, 68, 82, 93, 110, 114, 127, 133, 154, 155, 167
thixoforming, 187, 188, 189, 190, 214, 216, 217, 219, 221

U

Unglazed Plastic Collectors, 229

W

welding, 127, 128, 129, 130, 134, 144, 145, 148, 150, 234

Z

Zinc, 16, 243, 253
zinc stearate, 115, 116, 119, 123, 124